THE
CLASSIC
HORTICULTURIST

THE
CLASSIC
HORTICULTURIST

URSULA BUCHAN
AND
NIGEL COLBORN

CASSELL

LONDON

A CASSELL BOOK

Copyright © Quarto Publishing plc 1987

First published in Great Britain in 1987 by
Cassell Publishers Ltd
Artillery House, Artillery Row
London SW1P 1RT

British Library Cataloguing in Publication Data
Buchan, Ursula   Colborn, Nigel
The classic horticulturist.
1. Horticulture —— History —— Dictionaries
I. Title
635′.09 SB317.58
ISBN 0-304-32154-0

This book was designed and produced by
Quarto Publishing plc
The Old Brewery, 6 Blundell Street
London N7 9BH

SENIOR EDITOR Polly Powell
ART EDITOR Vincent Murphy

EDITOR Paul Barnett
ASSISTANT EDITOR Joanna Bradshaw
DESIGNERS Peter Bridgewater, Bob Cocker

ILLUSTRATORS Kate Osborne, Norman Bancroft-Hunt,
Marilyn Leader

PICTURE RESEARCH Anne-Marie Erlich

ART DIRECTOR Moira Clinch
EDITORIAL DIRECTOR Carolyn King

Typeset by QV Typesetting Ltd
Manufactured in Hong Kong by Regent Publishing
Services Ltd
Printed by Leefung Asco Printers Ltd, Hong Kong

Quarto Publishing plc would like to take this opportunity
to thank Jacky Morley and Andrew Kidd.

CONTENTS

INTRODUCTION   6

THE HORTICULTURISTS by Nigel Colborn   8
THE MIDDLE AGES   12
THE RENAISSANCE   14
THE 17TH CENTURY   18
THE 18TH CENTURY   29
THE 19TH CENTURY   35
MODERN TIMES   39
CONTINUING THE GARDEN TRADITION   50

A-Z DIRECTORY OF PLANTS by Ursula Buchan   52
PLANT LIST   53

FOOTNOTES   154
SELECT BIBLIOGRAPHY   155
SPECIALIST NURSERIES   156
OTHER USEFUL ADDRESSES   156
A SELECTION OF GARDENS OPEN TO THE PUBLIC   157
INDEX   157
ACKNOWLEDGEMENTS   160

INTRODUCTION

WHENEVER you bite into a sweet, juicy apple, or select a bunch of grapes, or breathe in the perfume of a blowzy pink rose on a summer evening, do you ever wonder how it is that we are lucky enough to be able to enjoy such pleasures? The enormous wealth of flowers, fruits, herbs and other useful plants that we have inherited is the result not just of natural evolution but of thousands of years of selection, collection and improvement of an almost infinite number of wild plants.

For example, although we might pay a ludicrous price for a bottle of vintage claret, all the currency of the world could not buy the thousands of years of development of the grape — from the time of our neolithic ancestors, who began to train wild vines over posts, through the ancient Persians, Egyptians, Greeks and Romans, the medieval monks ... and so on through history until finally we have the present diversity of vines grown all over the world.

Horticulture — the art and science of gardening — was the first step towards civilization. When our stone-age forebears began to grow their food plants near to their homes, instead of merely gathering nuts and berries from the wild, the resulting abundance gave them more time for other activities — such as thinking. It must be reasonably safe to assume that they gathered seeds or plants from wild stocks, so it follows that they would have selected the best examples. In fact, the art of improving plant varieties must have begun some 10,000 years before the science of genetics emerged. Food production must have been the first objective, but it is possible that certain plants were valued for their supposed magical properties. Later, the healing arts depended on specific plants for herbal remedies and, as civilizations evolved, people began to enhance their places of recreation with plants. Remains of tools have been found in the Tigris valley suggesting that people may have been gardening as early as 5000BC.

Through the ages there have been certain people who, because of their particular abilities, have had considerable influence on the development of horticulture. Some were great plant collectors, others pioneered technical advances, and yet others developed artistry in the garden. To tell the stories of some of these classic horticulturists and their achievements is to tell the story of horticulture itself.

THE HORTICULTURISTS

*H*ORTICULTURE was being practised long before the Hanging Gardens of Babylon were built in about 600BC. The Assyrians collected plants for their temple gardens and there are detailed records of ancient Egyptians growing figs, pomegranates and grapes. The lush silts of the Nile valley, refertilized every year by the river's winter floods, must have provided rich harvests. Queen Hatshepsut was introducing incense trees into her new temple garden between 1500 and 1485BC, and, during the reign of Rameses III (1198-1167BC), trees and shrubs were grown in decorated containers. Egyptian gardens had a line and formality that undoubtedly influenced the Romans, and were in due course to be adopted by the Italians. Today, most good gardens have a great deal in common with those found at the source of Western civilization several thousand years ago.

On the other side of the world the Chinese, during the Ch'in and Han dynasties, were developing the art of landscaping. Their approach seems to have been naturalistic — they enhanced natural scenery with artificial mountains, rock gardens and pools. No doubt trees like *Acer davidii* (see page 58) and Buddleia (see page 70) were enjoyed both in gardens and in the wild. The idea was to create an area for quiet meditation. The poet Hsieh Ling-Yin (about AD410) wrote[1]:

I have banished all worldly care from my garden, ... I have planted roses in front of the windows, but beyond them appear the hills.

A GREEK HERBAL

Some of the earliest detailed works on horticulture appear in Greek and Roman literature. Dioscorides[2] (*fl.* 1st century) was a Greek practising herbal medicine at about the time that *The Acts of the Apostles* was being written. He was born in Cilicia, and had served in the Emperor Nero's army. He wrote a book about pharmacology, *De Materia Medica*, that was to be used for the next 1500 years. He described the medicinal properties of 600 plants, as well as a large number of animal products which, he claimed, had medicinal value. We know little else about him but, thanks to a horticultural scholar called John Goodyer, Dioscorides' herbal was translated into English in 1655. Goodyer wrote out all 4,540 quarto pages, but the manuscript remained unpublished until the 20th century.[2]

Parts of the work are hard to believe and many of the illustrations are so botched that they bear little resemblance to the original plants — it is likely that they were copied from earlier copies, rather than drawn from life. It must be remembered, too, that our picture of Dioscorides is seen through the eyes of a 17th-

century Hampshire gardener — albeit a well
educated one — whose concepts of botany and
medicine were quite different from ours. There
is a lot of folklore and many of the remedies are
fanciful, to say the least. For example:

> *The stones [? testicles] of*
> *hippopotamus, dried and ground up*
> *small is drank in wine against the bitings*
> *of serpents.*

And, on the properties of *Cyclamen*:

> *They say that if a woman great with*
> *childe doe goe over ye roote, that she doth*
> *make abortion, and being tied round her*
> *it doth hasten the birth.*

But there is a wealth of useful information
too. He describes how to preserve orris root
(*Iris florentina*) and discusses its virtues. He
recommends the North African form as being
the best and says of the rhizomes: "When they
grow old they will be worm-eaten, yet then
they smell the sweeter."

When it comes to naming plants, anyone
with the merest smattering of botanical knowledge will wince at his classification. Plants were
grouped according to their medicinal affinity.

White bryony is placed with the botanically unrelated *Cyclamen* because both have the same pharmaceutical properties. He has the goodness to provide all the known names (around the Middle East at any rate) for each plant. He lists 13 for *Cyclamen hederifolium* — a good example of why a single, internationally recognized name for each plant is such a necessity today.

Dioscorides also seems to have set the trend whereby illustrations show the whole plant, complete with roots, flowers and fruits,

A Sienese edition of Pliny's *Historia Naturalis* of c.1460, showing Pliny instructing a gardener.

on the same page. Although at times this inhibits accurate representation, it does make for economy of space and enables the relevant part of the plant to be shown.

PLEASURE GARDENS OF ROME

Information about Roman gardens can be found in the works of Pliny the Elder (AD23-79)

Roman wall painting from the House of Livia; an early use of *trompe-l'oeil*.

and his nephew Pliny the Younger (AD62-c.114). They were hardly horticulturists — indeed, it is most unlikely that they ever so much as lifted a hoe — but their copious literary output reveals much.

The elder Pliny (Gaius Plinius Secundus) was a top civil servant for the Emperor Vespasian, with whom he was on friendly terms. In spite of a heavy workload, he managed to write a phenomenal number of books. His *Natural History* runs to 37 "volumes", and is a kind of *Life on Earth* without the benefit of scientific observation or colour television! Much of the information is hopelessly wide of the mark, based as it is on stories from unreliable sources. There are reports, for example, of peculiar men *(Sciapodes)* who had such enormous feet that they used them as sunshades, of cherries grafted onto willows, of unicorns, of winged horses, and so on. But there is also some sound material. He describes how a *topiarius* (not exactly someone to do the topiary, but a man employed to train creepers over statues and to keep growth in the gardens tidy) could dwarf planes and conifers with a technique similar to *Bonsai*.[1]

Pliny the Younger (Gaius Plinius Caecilius Secundus) had a lot to say about the plants and gardens of his day. He describes the use of box and rosemary for hedges, trees such as mulberry and fig, and sweet violets — frequently used for underplanting because of their scent. There are details of topiary work (hedges shaped into animals), cypress walks, statues and fountains. Clearly, Roman gardening had much in common with the modern art, and many Roman plants are enjoyed to this day.[3]

THE MIDDLE AGES

WE know little about what kind of horti-culture was being practised in Europe between the decline of the Roman Empire and the Renaissance because records are scarce. The vision one has of that period is of war-mongering gentry with private armies slogging out land disputes, while starving peasants squatted in wattle huts, terrified that yet another Viking gang might pop over the North Sea for a bit more rape and pillage. This is an absurdly exaggerated picture, but there was obviously little time for the niceties of artistic plantsmanship.

We do know, however, that after William the Conqueror's invasion of Britain in 1066, the development of new monasteries helped to preserve some of the species the Romans left behind. Plants like Paeonies, Opium poppies and Christmas roses *(Helleborus niger)* (see page 100) which are commonplace today, were grown as medicinal herbs. The 12th-century writer Alexander Neckham (1157-1217) describes such monastic garde-ning in his *De Naturis Rerum*, although whether he gleaned from practical observation or lifted from earlier literature is not clear. Throughout the next few centuries, records exist of vineyards, of vegetable gardens growing such plants as leeks, carrots and garlic and of fruit orchards. However, it is difficult to find much detailed and reliable information.

By the 15th century, horticulture was making progress again. Between 1400 and 1440, the aptly named Mayster Ion Gardener wrote a treatise called *The Feate of Gardening*. Very little is known about Gardener himself, but it is clear from his work that he had good, working knowledge and was an original

An illustration of a medieval garden from the 15th-century *La Roman de la Rose*, showing peas, irises and double roses.

The right wing of the Wilton Diptych,
c.1395. Campion, liverwort, double roses
and violets are shown.

thinker. Other writers of the time were merely regurgitating the theories of ancient authorities like Pliny the Elder, who had himself written largely from hearsay.[3]

Ion Gardener's text, by contrast, is highly practical and concerns itself solely with the culture of plants for utility. Novel techniques like the grafting of pear trees onto hawthorn are covered — it is interesting to see that clay and hazel bark were used instead of wax and raffia.[3, 4] Instructions on when to sow various seeds are given, and there is a good deal about parsley — an important herb and root vegetable. Gardener knew all about the value of using manure, and describes how to build up soil fertility for growing saffron — a precious medieval commodity, since it took 4,000 crocus flowers to make an ounce.[3] The "Saferowne," he said, must be planted "only in

beddys y-made wel with dyng."[4]

Of the 97 plants described in *The Feate of Gardening*, 26 had been introduced from outside Britain. No doubt there were other exotics being grown in monastery gardens and in the great households, but the large-scale collecting of plants had yet to begin.

Although it deals with utilitarian horticulture, Gardener's book mentions both roses and Madonna lilies, (*Lilium candidum*) (see page 106). These two plants both had medicinal uses but it is hard not to believe that, in an age when flowers were beginning to appear in paintings, at least *some* plants were being grown purely for decoration.

THE RENAISSANCE

*B*Y the time Gardener was making his record of horticultural practices, at the beginning of the 15th century, the arts were undergoing something of a rebirth. Up to 1400, records of gardening and botany are scanty and unreliable. The Norman monks must certainly have brought their skills to England: they have been credited with introducing pinks, wallflowers and other plants grown for medicine. We can tell from paintings like the Wilton Diptych (*c.*1395) that certain garden flowers were in cultivation. There are double roses, violets and irises at the Virgin Mary's feet. European wildflowers — Sea catchfly and Liverwort (*Silene vulgaris* and *Hepatica*) — are also shown, and were certainly being grown in the monastery gardens. Although their primary use was probably medicinal, their decorative properties must have been appreciated for them to have been used in such an artistic context.

By the middle of the century, science, painting and music were developing in Europe as never before, but in England the Wars of the Roses were to drag on for another 30 years. However, with the Tudors, innovation became widespread. Sir Thomas More wrote *Utopia* (1516), in which gardens figure as places with attractive flowers and well tended fruit (More had fine gardens laid out for himself at Chelsea), and Cardinal Wolsey built Hampton Court.

THE FIRST BOTANIC GARDENS

In Italy, at this time the epicentre of the Renaissance, gardening was progressing at a greater pace than anywhere else in Christendom. One result of the reborn interest in science was the birth of the botanic garden, the idea springing from the renewed interest in ancient literature: the first botanic gardens in the world were in Pisa and Padua.[1] Elsewhere, plant collecting was to become fashionable, particularly among royalty.

It was into this period of growing interest in new plants that Jules Charles L'Ecluse (Clusius) was born. Clusius was a great botanist and plant collector. With his sound technical education, he was able to observe and record botanical details with unprecedented accuracy. We tend to associate him with tulips — one thinks of the graceful pink and white *Tulipa clusiana* (see page 150) — but he discovered and collected hundreds of different plant species. His description of "Rosa Mundi" (*R. gallica* 'versicolor') (see page 140) — a plant widely grown and enjoyed today — is the first recorded.

His life was hard and he did not enjoy especially good health.[3] His family were Protestants, then in the minority, so it is likely that they suffered for their religious beliefs. He was well read and widely travelled.

From 1565, he spent a year or so plant hunting in Spain and Portugal. This was interesting for him not just because he could study the native plants but also because the Spanish and Portuguese were at the time bringing exotic plants home from the New World. It is possible that potatoes were already being grown in Spain as early as 1565, a couple of decades before Raleigh grew them in England, and Clusius may have observed them there — although he does not refer to them until much later, when he recalls in his *Rariorum Plantarum Historia* (1601) that he had the plant in 1588.[3]

On his visit to Britain (1571) he read Spanish physician Nicholas Monardes' book about American plants, *Joyfull Newes from the New Found Worlde*, and translated it into Latin (the *lingua franca* of 16th-century scholars) for use in Europe. (Frampton's English translation was not published until 1596.)

In 1573 he laid out a physic garden for Emperor Maximilian II of Austria at Vienna. He spent three years there studying the flora of Austria and Hungary, writing up all his findings in *Historia stirpium per Pannonium*, which was published in 1583. From Austria he moved to Frankfurt where in 1592 he was crippled by a serious fall. In spite of his failing health, he was persuaded by an ardent admirer, Johan van Hoghelande, to accept the post of Director of the Leyden Botanic Garden (part of Leyden University). In this appointment, at

---

JULES CHARLES L'ECLUSE (CLUSIUS)

Father of the Dutch Bulb Industry

−1526−
Born in Arras (then Flemish); later studied law and medicine at Wittenberg and Montpellier

−1565−
Trip to Spain: 200 species recorded

−1566−
Translated d'Orta's Indian Plants

−1571−
Visited England

−1573−
Laid out physic garden for Habsburgs. Studied flora of Austria– Hungary for three years

−1581−
Revisited England. Met Drake

−1587−
Dutch school of gardening formed

−1592−
Serious fall disables him

−1601−
Published Rariorum Plantarum Historia

−1609−
Became Director of Leyden (Hortus Botanicus Lugduni– Batavorum). Died

---

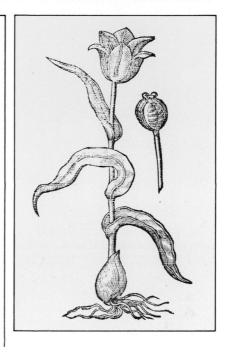

Tulipa clusiana, named after Clusius. This Asian species has been cultivated in Europe for centuries.

---

the end of his life, he introduced some of the first tulips, irises and crocuses. It is thought by some that these became the basis for the development of commercial bulb-production in Holland.

REBIRTH OF THE ROMAN GARDEN

While the botanists were busy collecting and classifying, the Netherlanders were improving cultivational techniques. When the Flemish weavers began to migrate to East Anglia, they brought many of their gardening skills with them. In England, the rich and powerful were having elegant gardens laid out around their houses. The link with ancient Rome was strong,

The frontispiece from Gerard's *Herball*, as originally printed in 1597.

JOHN GERARD

Genius or Plagiarist?

—1545—
Born at Nantwich, Cheshire; later educated at Willaston, near Nantwich

—1577—
Becomes superintendent of Burleigh's gardens in London and Hertfordshire

—1595—
Elected assistant to Barber—Surgeons' Company

—1596—
Published Catalogus arborum fruticum

—1597—
Appointed Junior Warden, Barber—Surgeons' Company and (December) published his Herball, or Generall Historie of Plantes

—1608—
Appointed Master of Barber—Surgeons' Company

—1611—
Died and was buried at St Andrew's, Holborn, London

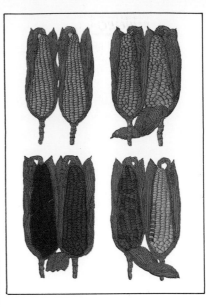

Maize (Indian corn), as illustrated in Gerard's *Herball*.

thanks to the Italian renaissance of classical styles. Gardens were formal and highly geometrical, with clipped hedges. Henry VIII had an artificial mound made at Hampton Court which was planted up with trees and shrubs.

In Elizabeth I's reign the knot garden developed. This consisted of a small area of convoluted hedgery with paths or beds in between. Hedging material was usually box or rosemary, and the beds were filled with coloured gravels or with flowering plants. Knot gardens were not very large — certainly not as large as the impressive French-style parterres.

Small gardens, although they probably contained some plants for beauty, were mainly functional. Thomas Tusser (1515-1580) wrote

Five Hundreth Pointes of Good Husbandrie (1573) which, in verse, describes how the house-wife was landed with all the gardening chores:

In Marche and in Aprill,
from morning to night:
in sowing and setting,
good huswives delight,
To have in their garden or
some other plot:
to trim up their house, and
to furnish their pot
Have millons [melons] at
Michaelmas, parsneps in Lent:
in June, buttred beans,
saveth fish to be spent
With these and good
pottage inough having than:
thou winnest the heart of
thy laboring man.

SURGEON GARDENERS

The Elizabethans' interest in herbs of all kinds was being stimulated by a succession of new discoveries, particularly from the New World. Surgeons of the day were keen to investigate the pharmaceutical properties of many of the new plants.

A prominent member of the Barber-Surgeons' Company was one John Gerard. Over the years, Gerard has come in for more than his fair share of lambast. Criticism of his *Herball* is well justified on many counts, but it does contain some original work and is an important example of Elizabethan botanical literature. His personality and humour radiate from the text.

Like his rather more famous contemporary, Shakespeare, Gerard came from fairly humble origins. Little is known about his early life, but it is likely that he travelled, probably as a ship's surgeon, before settling down to a London-based career. He married and, presumably, his wife helped him with his work — portions of the *Herball* are "mainly for women".[3] His reputation as a plantsman was well established by the time he was in his early thirties, and he was made superintendent of Lord Burleigh's gardens in London and at Theobalds, Hertfordshire. His own London garden held a collection of more than 1000 varieties, including rarities like double-flowered peach and white thyme.[8] He published details of this collection in *Catalogus arborum fruticum*, which was the first comprehensive catalogue of a private garden ever printed.

His first appointment as an officer in the Company of Barber-Surgeons came when he was 50. (He was steadily promoted until he became Master a few years before his death.) The *Herball* was published in the same year, 1597. The first edition, in spite of being a runaway bestseller, was riddled with mistakes. Some exotics were marked down as British natives and, worse, Gerard had stolen most of the original work from a Belgian botanist called Rembert Dodoens, failing to acknowledge his source. To illustrate the *Herball*, his publisher, John Norton, acquired a job lot of woodcuts from Frankfurt, which in several cases he matched to the wrong text.[1, 3] Having cobbled together the miscellany, Gerard added some of his own material and, to secure a stamp of quality and approval, dedicated the book to his patron, Lord Burleigh. The *Herball*'s popularity has remained undiminished for centuries.

One of its most interesting original illustrations is the portrait of Gerard himself. He is depicted holding a potato plant bearing flowers and fruit — the earliest known British picture of a potato. Love Apples (tomatoes) are described too. Early varieties were yellow, hence the Italian *pomadoro*. He says they should be

> *sown in a bed of hot horse dung after the maner of muske melons and such like cold fruits.*

Gerard's descriptions are concise and clear. Any remarks venturing an opinion are to the point. Of Sweet Williams *(Dianthus barbatus)* (see page 88), for instance, he says:

> *These plants are kept and maintained in gardens more to please the eye, than either the nose or belly.*

Although historians and botanists accuse Gerard of roguery, few condemn him outright. The *Herball* contains much of merit and its popularity cannot be denied. In the 16th century, copying other people's work was common practice. After all, at the same time, Shakespeare was penning *Julius Caesar* just around the corner from the Barber-Surgeons' Hall. The Bard's material was lifted from Plutarch, without acknowledgement, so who was the worse plagiarist?

THE
17TH CENTURY

*B*Y the time of Elizabeth I's death, the art of gardening in Europe was well developed. Several of her political counsellors, especially the Cecil family, had built grand homes for themselves and were keen to lay out new areas with flowers and fruit as well as to develop parkland. Knot gardens were incorporated into grand terraces which might be lined with walls and stairs or walks of clipped trees. Fountains and waterworks were becoming popular. At Hatfield House these were designed by Chaundler and engineered by a Dutchman — Sturtevant.[3] The essay *Of Gardens* (1625) by Francis Bacon (1561-1626) demonstrates how highly regarded good gardens were becoming in that century:

> *It is the purest of human pleasures; it is the greatest refreshment to the spirits of man; without which buildings and palaces are but gross handyworks.*

On a smaller scale, more people than ever were beginning to indulge in horticulture. Farmers were building larger houses and developing gardens that would provide a more interesting variety of fruit and vegetables. Better textbooks were needed, for the writings of Gerard and Thomas Tusser were becoming dated. Various authors produced helpful volumes, but the major work of the age was written by John Parkinson.

EDEN REVISITED

Parkinson combined three important qualities to create his masterpiece on English gardening, *Paradisi in Sole Paradisus Terrestris*. First, his education and training: as an apothecary he would have developed a detailed knowledge of all the plants used in his day. Second, he was an expert practical gardener who developed his own garden at Longacre and shared his expertise with contemporary botanists like John Tradescant the Elder. Third, he had, beyond scientific knowledge, a profound love and understanding of plants. These qualities, coupled with a talent for fluent and accurate writing, enabled him to produce a book which is both erudite and entertaining to read.

The title is Latin for "A park in the sun — an earthly paradise", a pun on his own name. The frontispiece alone could keep a keen plant historian amused for hours, with all the exotic plants of a 17th-century Eden — including pineapple and prickly pear — in the main body and the disclaimer (in French) at the bottom of the page:

> *Whoever wishes to compare art with nature and our parks with Eden, indiscreetly measures the stride of an elephant by the stride of the mite and the flight of the eagle with that of the gnat.*

The frontispiece from Parkinson's *Paradisi in Sole Paradisus Terrestris*, showing a 17th-century Eden.

JOHN PARKINSON

An Earthly Paradise

—1567—
*Born in
Nottinghamshire*[5]

—Before 1625—
*Became apothecary to
James I*

—1629—
Published Paradisi in
Sole Paradisus Terrestris*;
became Royal Botanist
to Charles I*

—1640—
Published Theatrum
Botanicum

—1650—
*Died and was buried at
St Martins-in-the-Fields,
London*

A page from Parkinson's *Paradisi in Sole Paradisus Terrestris* showing members of the family Boraginaceae.

There are nine chapters on practical gardening. The size and design of knot gardens is covered in some depth and there are some sound details about materials for their hedging. Cotton lavender, he says, will "perish in some places, especially if you doe not strike or put off the snow, before the sunne lying upon it dissolve it". There is plenty of good advice on siting a new garden, all of it readable and as topical now as it was 350 years ago.

The book goes on to describe a vast range of flowers, fruits and vegetables. The superb illustrations, by the father-and-son team of Christoph and Christoph Switzer, depict many of the var-

ieties in fine detail. There are double daffodils, 49 kinds of carnation and double pink, several species of iris ('Fleur de luce'), crown imperials, and so on. Further sections deal with vegetables and fruit, some of whose names are too delightful not to repeat: "The Towne Crab", "The Crowes Egge Apple" and the "Paradise Apple". In the fruit section there is a little eulogy on the benefits of (? fermented) apple juice:

*The juice of apples ... is of
very good use in
melancholicke diseases,
helping to procure mirth, and
to expell heaviness.*

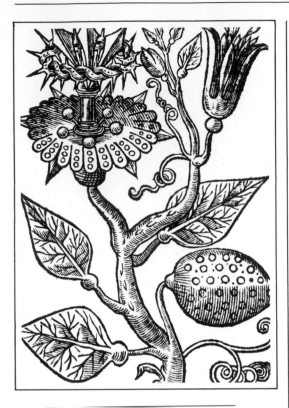

From Parkinson's *Paradisi in Sole Paradisus Terrestris*, an illustration depicting the Jesuits' idea of the Passion flower (*Passiflora*).

In 1640 Parkinson published a complete herbal entitled *Theatrum Botanicum*. Most of the material for this book — unlike *Paradisi*, which was based on his own observations — came from the work of a famous French physician and gardener, Matthias de l'Obel (after whom Lobelias are named).

As a plantsman, Parkinson was much respected and his contribution to gardening in the 17th century is incalculable. However, although he was a great chronicler and cohesive writer, many of the plants he described were collected by others, particularly the Tradescants. He demonstrated how exotic flowers could be used, but it took the obsessive collecting mania of the Tradescant family to help to enrich the world's heritage of garden plants.

STOCKING THE ARK

The lives of the two John Tradescants, father and son, ran from the glorious days of Queen Elizabeth through to the first couple of years of the Interregnum under Oliver Cromwell. Politically, the first two Stuart kings were not a great success but, despite the grim historical events of the era, science and the arts enjoyed tremendous patronage under them. The age began with Shakespeare's last and finest dramas, and went on to produce Inigo Jones, Van Dyck and Rubens. Royal interest in botany, after a strong beginning in Tudor times, was gathering momentum. Ocean transport was becoming safer and the quest was on for new fruits, flowers and vegetables.

John Tradescant the Elder was the son of a Suffolk yeoman who had moved to London.[9, 10] He did not marry until he was 37 and his son was born a year later in 1608.

His first recorded job was with Lord Salisbury, whose gardens were in London and at Hatfield House, some miles north. Hatfield was being revamped, and Tradescant was sent to find new trees for the avenues, fruits and vegetables for the orchards, and flowers for the "pleasure gardens". He introduced new grapevines, White mulberry — in the hope of starting silk production in England — and a number of new roses from France. By 1614 he had left Hatfield and moved to Kent, where he worked as gardener to Sir Edward Wotton.

A few years later, sponsored by Salisbury, he joined Sir Dudley Digges on what turned out to be a most fruitful expedition to Russia. Among many new introductions he brought back seeds of the first larch trees to be grown in Britain. Lord Salisbury had anticipated the expensive nature of the purchasing visit and had made special arrangements with the Treasury to provide bills of exchange, backed by his personal guarantee; these bills served as a kind of traveller's cheque. It is as well that

Tradescant was well financed for, by today's values, his purchases in Russia topped £25,000.

An obsessive nature, acute powers of observation and an indefatigable spirit — as with so many great horticulturists — spurred Tradescant Senior to greater achievements than those of any ordinary gardener. Descriptions of his travels, some of which are on record in his diaries, throw considerable light on his character.[9] It seems he was a great one for taking charge. On the voyage to Russia he nursed the seasick Digges. He observed whales, he caught migrating birds:

> *There were many small birds cam abord the shipe ... I have thre of their skins which were caught by myself.*

One remark suggests — and how unfair of Fate to such a great plantsman if the suggestion is true — that he lacked a sense of smell. He describes a patch of roses:

> *... much like our sinoment [cinnamon] rose; and those that have sense of smelling say they be marvellous sweete.*

He was a great improver, always looking out for better varieties and strains. He brought blackcurrants back from Eastern Europe because they were better flavoured than any in England at the time. The strawberries he left behind. They were

> *nothing differing from ours, but only les, which mad me that I did not so muche seek after them.*

THE TRADESCANTS
Royal Plant Collectors

*JOHN TRADESCANT
THE ELDER*

—1570—
Born in Suffolk

—1607—
Married at Meopham, Kent

—?—1614—
Worked for the Cecil family in London and Hatfield

—1618—
First major plant—collecting trip (to Russia)

—1623—
Took over Duke of Buckingham's garden at Newhall, Essex; made Keeper of Royal Gardens at Oatlands

—1626—
Moved to Lambeth

—1637—
Made Keeper of Oxford Botanic Garden, but too ill to take up the appointment

—1638—
Died

The gardens at Hatfield House, more than three centuries after John Tradescant the Elder helped to develop them.

After supervising the Duke of Buckingham's garden in Essex, Tradescant took a house at Lambeth which, because of the huge collection he had developed, became known as The Ark.

PLANTS FROM THE NEW WORLD

By the time John Tradescant the Elder had accepted the job of running the Duke of Buckingham's garden, his talented son, John Tradescant the Younger, was ready to leave King's School, Canterbury, and start an equally successful career in horticulture. No doubt the young enthusiast received the best training possible from his father. When the latter died in 1638, John the Younger succeeded him as Keeper of His Majesty's (Charles I's) Gardens at Oatlands.

The most important events of his career were his three trips to North America between 1637 and 1654. He was responsible for providing us with so many species which today are commonplace that it is hard for us to imagine what life was like without them.

The collection of plants and other artefacts at The Ark had become a famous attraction and, after his father's death, John the Younger set about drawing up a catalogue. This he published in 1654, making it the first catalogue of a private museum ever to be printed.

He had intended to leave the collection to the Crown on his death, but a crafty solicitor, Elias Ashmole, claiming friendship, managed to acquire it for himself. Ashmole bequeathed it to Oxford University, without due acknowledgement to the Tradescant family, where it became the basis of the Ashmolean Museum.[9, 10] Today, the importance of the Tradescants' posthumous contribution to the Ashmolean is at last recognized and acknowledged.

The plants discovered and introduced by the Tradescants are far too numerous to list, but a few examples will serve to show what a huge contribution the two men made. Parkinson mentions some of them in his *Paradisi in Sole Parad-*

JOHN TRADESCANT
THE YOUNGER

—1608—
Born at Meopham, Kent

—1623—
Started education at
King's School,
Canterbury

—1627—
Married Jane Hurte,
who died in 1635

—1634—
Made Freeman of
Gardeners' Company

—1638—
Appointed Keeper of
Royal Gardens at
Oatlands, Weybridge

—1638—
Married Hester Pooks

—1637, 1642, 1654—
Plant collecting trips to
Virginia

—1654—
Published Musaeum
Tradescantianum

—1662—
Died

isus Terrestris. There are pictures of double daffodils such as "Tradescant's Great Rose Daffodil", double martagons, double *Hepatica* and others. London plane trees (see page 126) were also bred from a Tradescant introduction; *Cotinus cogyggria* (Smoke bush, lilac), *Amelanchier ovalis* and *Smilacina racemosa* were theirs. They grew runner beans — as ornamentals, not realizing the food value of the green pods — as well as *Tradescantia* and *Juglans cinerea* (Butternut).

They introduced many new, improved varieties of fruit. Grapevines were brought to Hatfield House together with two new kinds of cherry: 'Archduke's' and 'Biggandre'. During his first visit to Paris Tradescant Senior developed a close relationship with Vespasien Robin, son of Jean Robin, gardener to Louis XIII of France — a friendship which resulted in many valuable plant introductions for Lord Salisbury.[9]

Few families have contributed as much to gardening as the Tradescants. The epitaph on the family tombstone in the graveyard of St. Mary's Church in Lambeth, London, sums up the extent of their explorative natures.

Know, stranger, ere thou pass, beneath this stone
Lye John Tradescant, grandsire, father, son,
The last dy'd in his spring, the other two
Liv'd till they had travell'd Orb and Nature through.

THE RESTORATION

With the end of Cromwell and the restoration of Charles II, the "Merry Monarch", society began to open up. The trend set by the king of going horse-racing, frequenting the theatre and leading a more permissive life than had been acceptable a generation before, was followed by his nobility. It is easy to forget that during this frivolous time Charles also founded the Royal Society to foster science; over 300 years later it is still going strong.

In Restoration gardens French influence ruled — as it was to do for a century. This was the age of the Sun King — Louis XIV — and such creations as Versailles. The biggest single name in French garden design was that of André Le Nôtre (1613-1700). He took the Renaissance designs (themselves inspired by Roman models) and broadened them into huge parterres, symmetrical lines of trees and complicated tapestry bedding. It could be said that the principal difference between English and French gardening at the time was that, while the French subdued nature, the English tried to enhance it. The records of three 17th-century Englishmen — Hanmer, Evelyn and Rea — tell us a great deal about the English interpretation of gardening fashion.

Sir Thomas Hanmer began to write about his garden during the English Civil War, finishing his *Garden Book* in 1659 (although it was not to be published until 1933). The gardens of large houses in the early 17th century were usually laid out in the French style, with formal parterres — knot gardens carried to extremes — consisting of embroidered patterns made with gravels, low hedges and little beds. The hedges were grown into arabesques and curliques so that the overall impression was of a huge patterned carpet.

By the time Charles II came to the throne, the style was changing. Although French influence, as noted, was to dominate garden design

Tradescantia virginiana, one of the countless plants collected by John Tradescant the Younger during his three trips to North America.

for some time to come, the growing interest in plants for their own sake was beginning to necessitate a change in the principles. It is impossible to grow a miscellany of interesting plants in a parterre without ruining it, so it became necessary to provide areas where the special needs of the new style of gardening could be catered for.

The passion for "greens" was getting under way at this time. Greens were evergreen plants in large containers which could be stood out in summer but which were brought indoors for winter — hence the name "greenhouses". Oranges were used for this purpose, their aromatic evergreen foliage being prized for its

Agapanthus umbellatus from an 1800 issue of *Curtis' Botanical Magazine.*

SIR THOMAS HANMER

A Garden in Wales

—1612—
Born in Welsh borders

—1646—
Retired to garden in Flintshire to avoid Civil War

—1659—
Finished his Garden Book *(not published until 1933)*

—1679—
Died

brightness. Various kinds of holly and laurel were used, as well as bay, oleander and *Viburnum tinus* (see page 152).

Hanmer's favourite plants were bulbs, particularly tulips. He called them

> *the Queen of bulbous plants, whose flower is beautiful in its figure, and most rich and admirable in colour, and the wonderful variety of markings.*

He describes nerines too, calling them "narcissus of Japan" — odd, because they are South African natives. (Most gardeners were as ignorant of the origins of exotic plants then as they are today.)

Hanmer writes up details of his fruit collection in Flintshire, particularly of his pears and grapes, of which a wide variety was then available. He also gives details of the plants he purchased, and it is staggering to see how expensive nursery stock then was: clearly, plant buying was not for the masses. He mentions, in 1667, having paid two shillings each for peach trees, three shillings for nectarine trees, and eightpence for Gillyflower (double pink) roots.[3] This writer remembers his mother buying pinks at Romford Market for eightpence each in 1957. Evidently, nurserymen fared rather better during the Restoration.

EVELYN'S DIARIES

Hanmer was a flower and fruit man. The first love of John Evelyn was trees. As a forestry expert he wrote about trees and their uses both as ornaments and as sources of timber. Of all the characters in this brief history, he is the most public. Many know him for his diaries, which describe his milieu, although not half so well as do those of Samuel Pepys. But he was a keen and educated gardener with a particular knowledge of trees. His book *Sylva, or A Discourse of Forest Trees* (1664) was a standard work to be used for another 100 years.[3] He was the first to recommend planting trees for the specific purpose of supplying wood — to the Royal Navy. Strangely enough, despite the soundness of his advice, no government attempted to initiate tree-planting for the Navy until the Napoleon-ic Wars.

Evelyn was not deeply impressed by the beauty of the parterre. He found straight lines and symmetry less to his taste than natural views. He enjoyed open parkland or woodland with broad, generous rides running through it. These rides, he claimed, enabled one to enjoy the view but, at the same time, there were trees nearby to offer shelter from the weather and to act as cover for game. He was also keen on encouraging songbirds, which were able to nest in the sheltered rides. He said they were "never found in lofty woods where they are exposed to hawks and owls".

In 1664 he published his *Kalendarium Hortense* in which he gives useful details for each month, such as the signs of the Zodiac, times of sunrise, and hours of daylight. There are sections on jobs of the month and notes on which plants are in their prime.

Evelyn was the first writer to describe a heated greenhouse (at the Apothecaries' Garden, Chelsea). In a later publication he describes a design of his own. In an age without fungicides, poorly ventilated greenhouses could soon result in "greens" being wiped out by botrytis and other fungal diseases. Evelyn's patent system was based on heat exchange: the draught of the flue was used to evacuate stale air from the building while, at the same time,

The first nerines in Britain (*N. sarniensis*), were described by Hanmer in his *Garden Book*. Shown here is the similar, but more robust, *N. bowdenii*.

fresh, heated air was drawn in by the resulting negative pressure inside the house.

Evelyn's own garden at Sayes Court in Deptford, London, was well known, and he frequently describes it in his diary. In the hard winter of 1683-84 the Thames froze over and a whole street-system of booths and shops was set up on the ice. In that February he writes:

I went to Says Court ... where I found many of the Greens and rare plants utterly destroyed; The Oranges and Myrtils very sick, the Rosemary and Lawrell dead to all appearance, but the cypress like to endure it out.

An event took place in 1698 which serves well as an object lesson in how to choose one's guests. Evelyn's pride and joy at Sayes Court was his 400-foot-long (120m) holly hedge. When he lent his house to the Czar of Russia — Peter the Great — His Majesty thought it would be fun to vandalize the garden, and he almost managed to wreck the beautiful hedge.[3] Describing the strength and beauty of the holly, Evelyn says: "It mocks at the rudest assaults of weather, beasts, or hedgebreakers."

Evelyn's writings were prolific. He was a well known scholar and society man.

JOHN REA — GENTLEMAN GARDENER

John Rea was a less flamboyant character than Evelyn but his book,

JOHN EVELYN

Diarist

—1620—
Born at Wotton, near Dorking, Surrey

—1638—
Educated at Balliol College, Oxford

—1643—
Visited Europe

—1647—
Married Mary Browne in Paris

—1649—
Published the first of his three dozen books (a translation)

—1660—
After tacitly supporting Royalist cause, was well received by Charles II on his accession

—from 1662—
Served on Royal Commissions on civil matters

—1662—
Elected to Royal Society

—1664—
Published Sylva, or a Discourse of Forest Trees and Kalendarium Hortense

—1685—
James II acceded to throne; Evelyn appointed Commissioner of the Privy Seal

—1706—
Died and was buried at Wotton

By the 17th century, tulips were widely grown in Europe. *Tulipa clusiana* was a popular species.

based on his own experiences, provides a delightful portrait of 17th-century English gardening. In fact, very little is known about Rea the man: all we have is the only book he wrote, *Flora, Ceres et Pomona* (1665). He styled himself "Gentleman", was a close friend of Sir Thomas Hanmer, and his book is one of the best gardening works to come out of the Restoration period. He had what must have been a fascinating garden at Kinlet in Worcestershire. That he was a first-rate plantsman is evident from *Flora, Ceres et Pomona*, which is subtitled: *A complete Florilege, fur-*

nished with all requisites belonging to a florist. He dedicated this work to, among others, members of the Hanmer family.

By the latter half of the 17th century, so many new plants had been introduced that Parkinson's *Paradisi in Sole* of 1629 was going out of date. Rea felt that there were too few good gardens about and promoted the idea of gardening quite strongly:

> *Fair houses are more*
> *frequent than fine gardens;*
> *the first effected by artificers,*
> *the latter requiring more skill*
> *in their owner.*

He goes into considerable detail about how to lay out a garden, with advice on where to build walls, which bits to have near the house and so on. He suggests that size should be limited, for "large gardens are usually ill-furnished and ill-kept". A nobleman, he suggests, might want about 80 yards (73m) of fruit and 30 yards (27m) of flowers; a private gentleman — someone like himself, presumably — would be able to get by with only 40 yards (37m) of fruit.[3] On the subject of fruit, he discusses how to graft pears onto quince stocks.

His advice about flowers is interesting but not always to modern tastes. The idea of having several different varieties of rose grafted to the same standard stock sounds a little grotesque by today's standards but is no less so than, say, standard fuchsias. He was one of the first to discuss how plants asso-

JOHN REA
Gentleman Gardener
—?—
Born
—1665—
Published Flora, Ceres
and Pomona
—1681—
Died

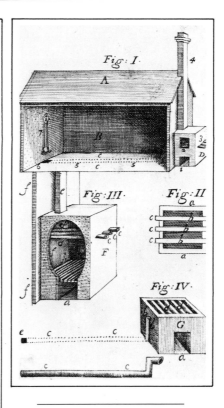

Evelyn's greenhouse allowed stale cool air to be exhausted while fresh warmed air was drawn in.

ciate with each other, and he describes a small bed with paeonies in the centre and dwarf anemones, ranunculi, tulips and irises round the outside. Gertrude Jekyll could have suggested the same combination a couple of hundred years later.

There are some helpful ideas on siting a summerhouse so that it can double as a bulb-sorting and -storing shed in the winter. He describes the service area of the garden in detail — the nursery, storehouses, hotbed and greenhouse. Rea shared Hanmer's love of tulips and their undeniable beauty even moved him to verse:[3]

> *The tulips to delight your eyes,*
> *With glorious garments, rich and new,*
> *Like the rich glutton some are dight*
> *In Tyrian-purple and fine white;*
> *And in bright crimson others shine*
> *Impal'd with white and graydeline*
> [*purple*]:
> *The meanest here you can behold,*
> *Is cloth'd in scarlet, Lac'd with gold.*
> *But then the queen of all delight*
> *Wears graydeline scarlet and white:*
> *So interwov'n and so plac'd,*
> *That all the others are disgrac'd*
> *When she appears and doth impart*
> *Her native beauties shaming art.*

Rea classifies his tulips according to their flowering period — early, middle and late — and describes nearly 200 varieties. He writes about a fair number of daffodils as well as molys *(Allium)*, asphodels, colchicums, gladioli and cyclamen. There are sections on auriculas, with special reference to particular varieties from places like the Oxford Botanic Garden.[3] The 1648 catalogue of the Oxford Botanic Garden contains details of a great number of rare primroses, including a blue one.[3] In the *Pomona* section of his masterpiece, Rea covers all the types of fruit we enjoy from modern orchards plus several, like *Sorbus* and medlars, which are rarely eaten nowadays. The familiar pear-names 'Chrétien' and 'Beurré' are listed.

Rea's is a book based on his own practical experience. He loved his garden and wrote about what he loved. He said: "It is knowledge that begets affection, and affection increaseth knowledge." He had no time for those who paid merely lip-service to plants. His dismissal of people who fail to recognize the value of a good plant is as apt now as in 1665:

> *I have known many Persons of Fortune pretend much affection to Flowers, but very unwilling to part with anything to purchase them; yet if obtained by begging, or perhaps by stealing, contented to give them entertainment.*

Oxford Botanic Gardens, visited by Linnaeus in 1735. Species of Mullein and Veronica can be seen.

THE 18TH CENTURY

*T*OWARDS the end of the 17th century, the increasing demand for plants presented commercial nurserymen with novel opportunities. The passion for standing hundreds of elegant pots about the place, each one planted with decorative "greens", was at its height. The number of species and varieties in cultivation grew bigger every year but, until this time, most nursery stock was imported, mainly from Holland. By the end of the century, George London had started to direct the first large professional nursery of its kind in England.

London was a self-made man of humble background.[3, 4] He was taken as pupil by John Rose, gardener to Charles II, who sent him to France, then the fount of gardening wisdom, to complete his apprenticeship. Around this time he was involved in the setting-up of the Brompton Park Nursery. The French sojourn was followed, in 1685, by a trip to Leyden, Holland, where he got to know a good deal about Dutch horticulture; the knowledge gained on these trips proved invaluable for his own nursery business later. On his return to England from Holland, he was appointed gardener to Bishop Compton, at Fulham, and proved himself to be exceptionally loyal and reliable. He became embroiled in the Glorious Revolution of 1688, when he was called upon by his master — and by Lady Marlborough — to assist in the escape of Princess (later Queen) Anne.

At about this time he teamed up with Henry Wise, and together they worked at a large number of gardens including Chatsworth, Blenheim and Windsor.[4]

By 1688, William and Mary were on the throne. George London was appointed superintendent of the Royal Gardens. In this post he was responsible for Hampton Court where, with Wise, he laid out "a great fountain garden" near Wren's new East Front. The famous maze was also their work; it was originally planted with hornbeam. Interestingly, the first recorded flowering Agapanthus was at this garden.

Hampton Court took up most of the royal budget for parks and gardens in London. In 1702, expenditure on plants and upkeep came to £1,623 out of a total budget of less than £2,000.[3] George London's annual fee for supervising the planting was £200. For a part-time job, this seems a princely sum, compared with a full-time gardener's weekly wage of four shillings, or a foreman's twelve shillings.

London's other great achievement had been, as we have noted, to set up the enormous Brompton Park Nursery at Kensington. The site is occupied today by the Natural History Museum, but in 1681 a consortium, headed by London, started the new business on 100 acres (40ha) — an ambitious plot by any standards. Other members of the consortium

The great semicircular garden at
Hampton Court, laid out by London and
Wise during the reign of William and
Mary.

were Moses Cook, John Field and Roger
Lookar. All were important gardeners —
Lookar, in particular, was gardener to Charles
II's wife, Catharine of Braganza. Henry Wise
joined the firm in 1689, and its name was
changed to "London & Wise".[3]

The name of the nursery's game was
"greens". A monumental greenhouse — bigger
than the king's own — was built, and production increased rapidly. By the turn of the century the firm employed a score of labourers and
held stocks of more than a million plants. By
the end of London's life, however, the craze for
containerized evergreens was beginning to
wane. Within a year of London's death Wise
sold up his share of the nursery, and soon afterwards it went into decline.

A DEDICATED PLANTSMAN

While the fashion for "greens" was fading, the
urge to collect plant species was still as strong
as ever. Botanic and physic gardens existed all
over Europe, and new introductions were con-

tinuing to pour into cultivation. Breeding, too,
was fast developing. The degree of success of
any garden depends on the calibre of the
people who run it. One British physic garden
achieved world fame largely because of the
energy of one of its early gardeners.

The story of Philip Miller is linked with the
development and success of the Chelsea
Physic Garden in London. Towards the end of
the 17th century, the Apothecaries' Company
had wanted to acquire some land outside
London on which to grow their collection of
medicinal plants. In 1673 they leased a parcel
of riverside ground from Charles Cheyne, very
close to where Sir Thomas More's estate had
been in Henry VIII's reign. The soil was light
and fertile and, once the enclosure had
been walled in, they had a sheltered spot
in which to grow their exotics. In 1722 Sir

Hans Sloane, himself a keen botanist, purchased all the land and conveyed it to the Apothecaries' Company, subject to certain conditions: the garden was to be properly managed and, for 40 years, 50 new species per year were to be grown in the garden. A dried, mounted specimen of each new plant was to be presented to the Royal Society. This ensured that at least 2,000 new plants would be introduced before 1762.[11] The apothecaries appointed a professional florist as gardener, and in Philip Miller they made a wise choice. He was to work at the Physic Garden for a staggering 48 years.

Miller, a Quaker with Scottish ancestry, was instilled with a deep commitment to hard work. His father had come south to set up a market garden in Kent, and so Miller would have grown up with the florist's art. (Florists were breeders and selectors of specific plants, not flower traders.) His *Gardeners' and Florists' Dictionary* (1724) became the standard reference work for florists for more than a century. His encyclopedic knowledge of plants and his practical background were helpful, but in addition to these qualities, he was lucky enough to be possessed of green fingers. There are scientists alive today who may be spearheading the frontiers of botany, but who need merely to glance at a geranium to have it wither and die. Others, without a scrap of formal training, can tear off bits of plant at the wrong time of year and still have them root in a

GEORGE LONDON

Nurseryman and Landscape Expert

—1640—
Born

—1681—
Partner in Brompton Park Nursery

—1685—
Visited Leyden

—1687—
Joined by Wise; worked with Wise at Chatsworth and elsewhere

—1688—
Glorious Revolution; assisted Princess Anne to escape; appointed superintendent of Royal Gardens for William and Mary

—1700—
Invited to design gardens at Castle Howard

—1713—
Died

The frontispiece to the *Gardener's Kalendar* for 1745, showing Flora welcoming the god of horticulture.

matter of days. Miller had both the training and the "way" with plants.

Hotbed culture — the heat generated by the decomposing material in the bed is supposed to protect the items planted in it — was a Miller speciality. Greenhouses and stove-houses were used for tropical plants, and one of the most celebrated trees in the collection, mentioned earlier by Evelyn, was the Jesuit Bark (quinine). The cedars of Lebanon at Chelsea became quite famous, too. They were not the first to be planted in England — John Watts, one of the early curators, planted them in 1683 — but they were the first in England to produce cones.[3]

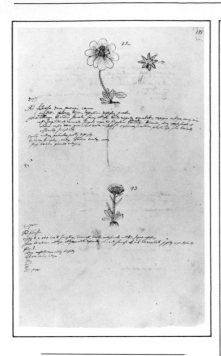

A page from Linnaeus' notebook, illustrating two Scandinavian natives, *Dryas* and *Erigeron*.

PHILIP MILLER

The Chelsea Physic
Garden

—1691—
Born

—1722—
Appointed gardener,
Apothecaries' Garden,
Chelsea (Chelsea Physic
Garden)

—1724—
Published Gardeners'
and Florists' Dictionary

—1736—
Linnaeus visited Chelsea
Physic Garden

—1771—
Died

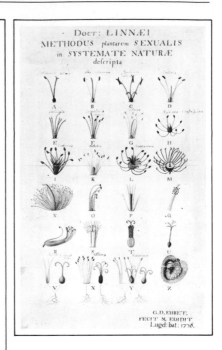

One of Linnaeus' illustrations of the morphology of flowering plants, showing the stamens and pistils of various species.

Miller received Linnaeus at Chelsea in 1736, and the fact that the great man visited at all shows how much the garden, with its extensive collection of rare plants, had grown in international stature since Miller's appointment. Miller, by 16 years the older man, disagreed with Linnaeus' new taxonomic theories until many years later, when he adopted them in revised versions of his *Dictionary.*

Miller, in his declining years, made the mistake of clinging too long to his appointment. The fame of the Chelsea garden became overshadowed by Kew which, under the vigorous management of William Aiton, became the botanist's Mecca. However, to this day the Chelsea Physic Garden remains an important repository of rare plants.

An example of how such places can serve mankind can be seen in the North American cotton story. According to Dr Dawtrey Drewitt,[11] writing in 1922, Miller sent a packet of upland cotton seed to Georgia in 1732. From that single original packet grew the North American cotton industry. A little hard to swallow perhaps, but not impossible.

Towards the end of the 18th century, French influence declined. English love of nature was breaking through the strictures of regimented planting. Parterres were going; broad, sweeping lawns were coming in. By 1760, work at Stourhead had begun, and with it came the fashion of constructing lakes large enough to reflect a topography decorated with trees and shrubs. The second half of the century saw the works of the great landscapers Lancelot "Capability" Brown and Humphry Repton.

THE NAMING OF PLANTS

Botanical science was moving forward rapidly, too. With so many new plants to classify, the world was crying out for someone clever enough to introduce a soundly based taxonomy. Such a man was the Swedish-born Carl von Linné (Carolus Linnaeus).

Before we complain about the awkwardness of Latin names for plants, we should ponder on how things were before an agreed international naming system was developed. With colloquial names confusion reigns. In England, for example, the name "Grannies' Bonnet" refers to: *Aquilegia vulgaris*, *Geum rivale* and *Silene alba*. If you talk to a Japanese botanist about "brandy bottles", you will confuse him, but he is likely to recognize the name *Nuphar lutea*. True, some scientific names seem unnecessarily difficult — who could forgive the boffins for *Paeonia mlokosewitschii*? — but at least they are standard throughout the world.

Botanists had attempted to find a satisfactory way of classifying plants for centuries, but it was Linnaeus who developed a system of naming pretty well everything that lived. He was not only a brilliant botanist but a first-rate doctor of medicine: he ran a successful practice in Stockholm, and developed a new cure for certain venereal diseases.[12]

His passion for plants and flowers began during childhood. His father, a curate at Smaland, had a plantsman's garden at his vicarage. After a year at Lund University, Linnaeus went to Uppsala, where he qualified and where he was offered a lectureship in botany at the

The Chelsea Physic Garden, made famous by Philip Miller. The statue in the centre is of Sir Hans Sloane.

youthful age of 23. In 1735 he travelled to Holland and took another medical degree, at Hardewijk. During this period he went to England to visit the famous Apothecaries' Garden at Chelsea where, as we saw, he met Miller. He went on to the Oxford Botanic[3] Garden but returned to Sweden in 1738.

There followed a period of practising medicine; during this time the University of Uppsala offered him their chair in medicine. His intellectual talents were, by now, phenomenal, and he was able to present lectures in zoology, botany, geology, medicine and hygiene; in addition, he was an accomplished writer and poet! He was appointed Royal Physician and, by 1758, had generated enough wealth to be able to buy a country estate at Hammerby, just outside Uppsala.[12] Recognition came from King Adolphus Frederick of Sweden, who honoured him with the title von Linné.

Linnaeus in the costume of a Laplander. His tour of Lapland in 1732 led to a detailed study of the flora.

CARL VON LINNÉ (CAROLUS LINNAEUS)

The Principle of Genera and Species

—1707—
Born May 23rd at South Rashult, Sweden

—1730—
Became lecturer in Botany at Uppsala University

—1732—
Visited Lapland

—1735—
Produced Systema naturae

1737—
Published Flora Lapponica *and* Genera plantarum

—1739—
Married

—1741—
Appointed to the Chair of Medicine at Uppsala University

—1742—
Exchanged his Chair in Medicine for that in Botany

—1753—
Published Species plantarum

—1761—
Ennobled, and granted the title von Linné

—1774—
Suffered serious stroke

—1778—
Died and was buried at Uppsala Cathedral

Linnaeus' most important works were the 1735 *Systema naturae*, the 1737 *Genera plantarum* and the 1753 *Species plantarum*. With the growing numbers of new species being discovered in the 18th century, the world was crying out for a reliable classification system. Linnaeus' plant nomenclature was scientifically based. Everything was grouped into genera and species; every plant had a binomial name. (*Genus*, beginning with a capital letter, and *species*, with a small letter — thus *Bellis perennis*.) Classification was based on the plant's anatomy. He used such parameters as numbers of stamens, numbers of pistils, and whether petals were joined or free, so that botanists could place any plant into a named category based on its natural characteristics. There was worldwide agreement that *Species plantarum* and the fifth edition (1754) of *Genera plantarum* together provided the starting point for international nomenclature.

As well as his work on taxonomy, Linnaeus produced some important reference works on botany. He explored Lapland in 1732 and wrote a detailed flora for that region which was published in 1737 and translated into English in 1811 by J.E. Smith. (Smith was first president of the Linnaean Society — a scientific club which he started with Sir Joseph Banks in 1788, when they managed to purchase Linnaeus' library and herbarium from Sweden.) A Swedish flora was published in 1745, followed by a fauna in 1746.

In 1755 the King of Spain invited Linnaeus to settle in Spain with a large salary. This tempting offer Linnaeus declined, preferring to continue his work in Sweden.

In 1774 he suffered a serious stroke, and from then onwards his health failed. He spent the last four years of his life in a state of semi-consciousness; his death in 1778 came as a release. He was buried in Uppsala Cathedral.

Linnaeus was a very great man with a giant intellect. Of him, King Gustav III of Sweden said:

> *I have lost a man who has done honour to his country as a loyal subject, as well as being renowned throughout the world.*[12]

THE
19TH CENTURY

*B*Y 1800 the English style of garden, as typified by places like Stourhead, was beginning to influence the world as strongly as had the French 100 years before. Areas colonized by the British were to inherit their style of gardening. Climate made a difference to the type of plants used, but not to the style. One of the finest examples in the tropics is at Peradeniya, Sri Lanka, started in 1821. Botanic gardens were laid out also in Singapore (1822), Trinidad (1819) and Jamaica (1774). One of the loveliest, not only for its layout with lawns, natural groups of trees and flowering plants, but also for its magnificent position in a quiet harbour cove, is at Sydney, Australia.

BEGINNINGS OF "VILLA" GARDENING

Small gardens, too, were beginning to develop. Changes in the social structure of the civilized world were to mean changes in gardening habits. One of the first visionaries to herald this new age was John Claudius Loudon.

Loudon was born within a few years of the death of Linnaeus. In Europe it was an ominous time: the French Revolution was about to shake the old establishment to its roots, the first convicts were being sent to Botany Bay, and George III was showing signs of lunacy. The Agricultural Revolution was progressing rapidly, soon to be followed by the dawn of modern industry.

Up to the 19th century gardening was either carried on by cottagers — more from necessity than for pleasure — or indulged in on a grand scale by wealthy landowners. By 1800, with changing fortunes, a number of newly successful individuals were building "villas" for themselves. By today's standards these properties would have been considered large — perhaps ten acres (4ha) of grounds and half a dozen principal bedrooms — but they were really small-scale imitations of the large landowners' mansions. A middle class was emerging too, and with it a taste for suburban life and suburban gardening. John Loudon was one of the main innovators in this new mass-gardening age.

An illustration from an 1832 edition of Loudon's *Gardeners' Magazine*, showing an early lawn mower.

He was born in 1783 near Edinburgh, where he went to school until he was 14. His thirst for knowledge of all kinds was prodigious. He was fascinated by scientific subjects, especially chemistry, and he was a capable draughtsman. Because of his love of plants, his father sent him as an apprentice to a local nurseryman, under whom he studied trees and their culture with such intensity that he soon became an authority. During his apprenticeship he taught himself French and Italian. In 1803 he went to London, intending to set himself up as a freelance journalist. He made friends with Sir Joseph Banks, who allowed him access to his private library and who, no doubt, encouraged his advancement. Certainly Banks assisted in Loudon's election into the Linnaean Society in 1806.

In 1813, after Napoleon's retreat from Moscow but before he had been defeated in France and packed off to Elba, Loudon took it into his head to travel to Russia. For a normal person, such a trip would be about as much fun as a sightseeing tour of Warsaw in 1942. But Loudon was tougher than most. The adventures that befell him on the way make better reading than any fiction: he was attacked by Cossacks, left alone in a snowdrift surrounded by howling wolves, and arrested as a spy.[13] In spite of the vicissitudes of the trip — Moscow was a smoking ruin — he managed to visit every famous garden and to make his mark on Russian society. He was elected a member of the

JOHN AND JANE LOUDON

First Steps to Popular Gardening

JOHN CLAUDIUS LOUDON

—1783—
Born April 8th in Lanarkshire; later educated in Edinburgh

—1797—
Apprenticed to a nurseryman; taught himself French, Italian

—1803—
Moved to London

—1806—
Elected to Linnaean Society

—1807—
Right arm amputated

—1813—
Visited Russia

—1822—
Published Encyclopaedia of Gardening

—1826—
Launched The Gardeners' Magazine

—1831—
Designed Birmingham Botanic Garden

—1830—
Married Jane Webb

—1836—
Launched The Suburban Gardener

—1838—
Published Arboretum et Fruticetum Britannicum

—1840—
Visited Europe

—1843—
Died (December 14th)

Imperial Society of Moscow. His impression of Russian nobility was not so favourable. He wrote:

A barbarous people may hang together by a sort of tattered moral principle ... the simple principle of self preservation.

As an excessively hard worker, it is not surprising that Loudon had managed to make himself pretty well off by the time he was 30. By 1812, through writing, farming as a tenant in Middlesex and Oxfordshire, and landscape consultancy, he was worth £15,000.[13] In 1823 he built himself a comfortable house in Porchester Terrace, Bayswater, about a mile from Marble Arch, London. It was typical of the sort of villa that successful men were building for themselves. In its tiny domed conservatory he built up a collection of Camellias. (Conservatories were yet to reach Victorian grandeur as seen today at Kew and at Longwood, Pennsylvania.)

His *Encyclopaedia of Gardening*, published in 1822, was aimed at people who, like himself, were fairly new to gracious living. The *Encyclopaedia* is a "how to" book of roughly 1.1 million words with helpful illustrations and descriptions. First comes a long section about the history of gardening, followed by details of tooling-up for the estate, man-management and plants. It has been compared with Mrs Beeton's *Household Management*, doing for small estate management what she did for housekeeping.

The gardens of the Royal Horticultural Society at Kensington. In 1822, the society moved to a larger site at Chiswick.

Overwork, carried to Loudon's feverish extremes, did nothing for his health. He suffered from rheumatics, particularly in his right arm. As a result of charlatan treatment, the arm was accidentally broken and amputation was advised. His behaviour at what must have been an excruciating operation was stoical to say the least. One of his draughtsmen was staying with him at the time and recorded the event:

> *After lunch he walked upstairs [for the operation] quite composedly, talking to the doctors on general subjects. When all the ligatures were tied, and everything was complete, he was about to step downstairs, as a matter of course, to go on with his business; and the doctors had great difficulty to prevail upon him to go to bed.*[13]

Rheumatism developed also in the other arm, so that he could move only his third and fourth fingers.

In 1826 he launched one of the first popular gardening magazines: *The Gardener's Magazine*. A couple of years later he came across a new, futuristic novel, *The Mummy! A Tale of the Twenty-Second Century*, by Jane Webb. It contained descriptions of 22nd-century mechanical ploughs and milking machines. He was so impressed that he wrote a review in *The Gardener's Magazine* and contrived to meet the author. They met in February 1830 and were married the following September. She became his right hand — almost literally! — taking dictation, helping him direct the gardeners, and fetching and carrying for him with total devotion.

BOTANIST BY MARRIAGE

At the time of her wedding, Jane Loudon said:

> *It is scarcely possible to find any person more completely ignorant of everything to botany and gardening, than I was at the period of my marriage with Mr Loudon ...*

But, like her husband, she had a hungry mind and was a keen worker. She began to attend the lectures of Professor John Lindley at University College, London, and, with expo-

An illustration from Jane Loudon's *Ladies' Flower Garden of Ornamental Perennials* (1844).

JANE WELLS LOUDON née WEBB

—1807—
Born August 19th in Birmingham

—1827—
Published The Mummy!

—1830—
Married John Loudon

—1831—
Went on tour of gardens in the North

—1840—
Published Gardening for Ladies

—1841—
Began to publish Ladies' Companion to the Flower Garden *series of books*

—1843—
Published The Entertaining Naturalist; *continued to publish further garden books until:*

—1852—
My Own Garden, *her last book*

—1858—
Died

sure to her husband's fund of knowledge, picked up a phenomenal amount of information in a very short time. The Loudons became a productive team. As his amanuensis, she helped him to complete his most important book, *Arboretum et Fruticetum Britannicum*, a massive undertaking in eight volumes, published in 1838.

On a trip to Birmingham, a few months after their wedding, the Loudons visited Chatsworth, in Derbyshire, where Joseph Paxton, one day to build the Crystal Palace and receive a knighthood, was head gardener. Although they became friends, Loudon and Paxton were natural opposites. Loudon disliked what Paxton was doing with the gardens at Chatsworth, where flamboyant bedding-out was practised. For Loudon, the beauty of individual plants mattered more than the

mass effect, which he found unpleasantly ostentatious. In his own garden in Bayswater he had a rich collection of more than 3,000 species. Every inch of ground was used yet, despite the profusion, the garden was well designed and pleasing to the eye. It seems ironic that, despite Loudon's botanical approach to villa gardening, the Victorians were to abandon plantsmanship and go for jazzy bedding out in its place.

By 1840 Loudon was ill and in debt. He was working neurotically hard, landscaping, designing cemeteries and writing, frequently through the night. This meant that Jane would have to write through the night as well, for it was at his dictation that she worked. She was also producing her own *Gardening for Ladies* (1840) and the *Ladies' Companion* series at this time. A crushing blow came when his major creditor went bankrupt. The wolves closed in on Loudon at a time when he was too ill to do much more than dictate to his wife. They managed to borrow some money from Joseph Strutt of Derby, but Loudon had to assign the copyright of his forthcoming book on Repton to his creditors. At dawn on December 14th, 1843, after an all-night dictating session, he collapsed in his wife's arms and died.

Jane Loudon continued to live at Porchester Terrace until her own death, 15 years later. She had become Britain's leading garden writer, and had published a considerable number of books about natural history and plants.

MODERN TIMES

THE Loudons' love of naturalism in landscapes and gardens, although it went out of fashion for a generation, set the trend for modern "informal" gardening. Although one can still find uncomfortable reminders of the Victorian bedding craze in public parks, corporation traffic islands and certain private gardens, much of today's popular gardening stems from Loudon's original philosophy, improved and interpreted by the great gardeners of the last 100 years.

In the modern age there was to be little more fancy landscaping on the lines of Brown and Repton. Much laying out and designing of small gardens was to be done and, more and more, the emphasis was on the beauty of the plants themselves. In the 1880s, the strong backlash against hard, formal planting was led by a gardener called William Robinson.

NO TO CARPET BEDDING

Born in Ireland in 1838, Robinson worked in a number of Irish gardens before crossing the sea in 1861 to take up work in Regents

Robinson's rose garden at Gravetye Manor. The planting is fairly relaxed by the standards of the early 20th century.

Robinson's specially converted Citroën. This enabled him to be driven to the less accessible parts of his Gravetye Estate.

WILLIAM ROBINSON

A Return to "Natural" Gardening

—1838—
Born in Ireland; later worked at various gardens, rising to the position of foreman

—1861—
Went to Regents Park, London

—1867—
Went to Paris as garden correspondent for The Times

—1868—
Visited the Alps

—1870—
Published Alpine Flowers for Gardens *and* The Wild Garden; *visited North America*

—1872—
The Garden *magazine launched*

—1879—
Gardening Illustrated *magazine launched*

—1883—
Gravetye Manor purchased. The English Flower Garden *published; further publications over next 40 years*

—1909—
Disabled by a bad fall

—1935—
Died

❋

Park, London. By this time, the mania for carpet bedding was at its height in England. Encouraged by trend setters like Joseph Paxton, ostentation and artificiality ruled. Gardens were laid out in great wodges of lurid colours, punctuated by bushes clipped into peculiar shapes and surrounded by banks of depressing evergreens. Obelisks, urns and curiosities abounded. Lingering in the garden was not enjoyable. It hurt your eyes and raddled your senses.

Robinson was a lover of wild flowers and of trees. In his lifetime he was to plant hundreds of thousands of them.[15] He was also to set the trend towards naturalistic planting. He did not go all the way — indeed, by later standards, some of his planting was formalized — but he did begin the revolt against a form of gardening that was contrived and selfconscious.

His origins were humble[4]. He probably received little education but, by the time he moved to England, in his early twenties, he was an experienced foreman. At Regents Park he was put in charge of the herbaceous section, and was promoted rapidly. He began to collect wild flowers and, on country excursions, became familiar with the English cottages. He delighted in the jumble of plants in their gardens — herbs, fruit and flowers all apparently planted without any special design and yet all with specific uses. His observations inspired him to begin a career of garden writing.

He had provided *The Gardener's Chronicle* with several articles by 1867 when *The Times* sent him to Paris as a special correspondent. He wrote about French parks and gardens and later, after a tour of the Alps, published *Alpine Flowers for Gardens* (1870). More books followed quickly: *The Wild Garden* (1870); *Hardy Flowers* (1870) and *The Subtropical Garden* (1871).

Robinson's early work was not heavily critical of "formal" French parkmanship. However, as the years went by, his output became more abrasive until it developed into a crusade against what he called "straitlaced" and "mechanical", and in favour of "natural" style. He wrote:

> *I am a flower gardener, and not a mere spreader about of bad carpets done in reluctant flowers.*

In this respect Robinson and the Loudons had much in common. Loudon wrote copious notes about hedgerow flowers and cultivated florists' varieties in his garden, detesting what he called "tawdry" displays.

Robinson began several magazines. In 1872, after a trip to North America, he launched *The Garden*, which was financially disappointing but enabled him to further his crusade. *Gardening Illustrated* began in 1879 and was. to run until 1956, when it merged with *Gardener's Chronicle*. Several other periodicals were produced during his life. In 1883 his most important work, *The English Flower Garden*, was published; in revised editions, this remained in print for about 80 years. It promoted "natural" gardening:

> *One aim of this book is to uproot the idea that a flower garden must always be of set pattern on one side of the house.*

By now he was becoming one of the most influential voices in British horticulture. With his ill disguised attacks on individuals and their gardens, he was making enemies too. Even dead champions of the age were not exempt. He called Paxton's Crystal Palace garden "the fruit of a poor ambition to outdo another ugly extravagance — Versailles".[4]

By 1883 he was rich enough to buy a country estate, Gravetye Manor, Sussex. Here at last, after so many years of working in other people's gardens, he was able to put into practice much of his gardening philosophy. At Gravetye he created his natural garden. He used herbaceous perennials such as *Aster novae angliae* (see page 68) not only in borders but naturalized in grass among shrubs. He is usually credited with being the first to plant bulbs in grass, but Loudon did this in his lawn in Bayswater,[13] and it is hard to believe that grass in earlier years had not been studded with at least bluebells and fritillaries, if nothing else. Robinson was a lover of small plants —

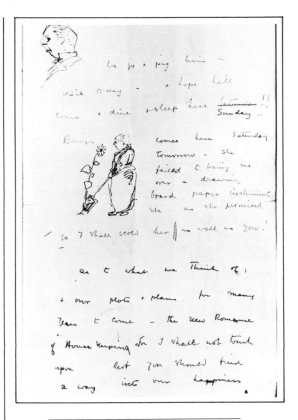

From a letter by Phipps, about 1906, a cartoon sketch of Gertrude Jekyll — nicknamed "Bumps" by Sir Edwin Lutyens.

Viola tricolor hybrids, walls planted with *Erinus* and yellow fumitory. In the fourth edition of *The Wild Garden* (1894) he describes a white *Clematis* growing up into a yew tree, a carpet of sweet cicely growing among shrubs, and roses growing into trees.

In 1909, at the age of 71, Robinson injured his back and was disabled. He continued to write and to travel about his estate in a specially converted vehicle.

THE GREAT "MISS BUMPS"

Robinson may have restarted the trend to naturalistic gardening some 30 years after Loudon's death, but it was Gertrude Jekyll who perfected the style. In fact, Robinson and Jekyll were

The gardens at Hestercombe, Somerset —
another product of the Jekyll-Lutyens
partnership.

friends for many years and worked together on *The Garden*, which she edited for a short time when he stood down. There is no doubt that the two horticulturists influenced each other, although it is not clear who had the greater effect upon whom.

It is fashionable today to rank Jekyll as one of the greatest figures in the history of horticulture. That she has had a profound influence on modern gardening is indisputable. Her colour schemes, her plant associations, her treatment of wild gardens and her approach to garden layout can be seen everywhere. Her long association with Sir Edwin Lutyens, the architect, is hailed by many as a triumphant coming together of ideas. Looking at it another way, one could suggest that Jekyll managed some pretty successful planting schemes in spite of an excess of garden architecture. Without doubt, her sense of artistry shone through her plant groups regardless of whether they were in a constructed garden or in a tract of unspoilt woodland.

Her parents were wealthy and respected but had no time for the strictures of the Victorian age. They allowed her to attend art school (female students were rare), where she learned to become a competent artist and craftsman. She managed to get a painting into the Royal Academy and, encouraged by a good friendship with William Morris, began to work on embroidery, woodcarving and even silverwork.[16]

She travelled extensively, and assisted with interiors for such places as Girton College and the Duke of Westminster's Eaton Hall. Besides Morris, her friends included Ruskin, Burne Jones and Brabazon — who taught her some of her colour sense. She was evidently a member of the artistic set. She never married, and because of her tubby stature was nicknamed "Bumps" by Edwin Lutyens.

Poor eyesight was a serious weakness. She must have known that her eyes were deteriorating, because in 1891 she consulted a Dr Pagenstecher in Wiesbaden, who announced what must have sounded like a heavy penal sentence. She was told that, unless she stopped all close work — all embroidery and painting — her sight would fail. Although this did not mean financial ruin, it did prevent her from pursuing her career.

It was at this point in her life that she turned her attention to gardening. She began to write copiously — history fails to relate why it was safe for her to write but not to paint or embroider! — and to design planting schemes. By now her partnership with Lutyens was developing.

Jekyll's genius lay in her ability to place the right plants together. She was able to combine their different characteristics — not only their colours — to create an artistic whole, much as any artist makes use of the materials at his or her disposal. To do this successfully is more difficult than it sounds, and even in the best gardens there are near misses; elsewhere there are disasters. In her *Colour in the Flower Garden* (1908) she wrote:

> *Having got the plants, the great thing is to use them with careful selection and definite intention. Merely having them planted unassorted in garden spaces, is only like having a box of paints from the best colourman.*

GERTRUDE JEKYLL

Painter turned Gardener

−1843−
Born November 29th in London

−1848−
Moved to Bramley, Surrey

−1861−
Entered Kensington School of Art

−1866−
Had painting accepted for Summer Exhibition, Royal Academy of Arts

−1868−
Moved to Wargrave Hill, Berkshire

−1875−
Association with William Robinson began

−1876−
Father died. Moved to Munstead Heath

−1889−
Association with Sir Edwin Lutyens began

−1891−
Crisis with her eyesight

−1896−
New house at Munstead Wood begun

−1900−
Published Home and Garden, *to be followed by many more books over the years*

−1908−
Published Colour in the Flower Garden

−1932−
Died (December 8th)

In her own garden at Munstead Wood in Surrey she was careful to retain as much woodland as possible, thinning trees here and there until groups and walks were left. She planted the shaded areas with woodland plants — hellebores, primroses, anemones and bulbs. Her 'Munstead Bunch' primroses (a range of polyanthuses from white to deep gold) were developed, not by a technical plant-breeding programme, but simply by collecting and sowing seed from the best plants.

THE LUTYENS/JEKYLL PARTNERSHIP

The value of her association with Lutyens is difficult to assess. In gardening there are two extremes of opinion. On the one side, the plantsman argues that a garden's most important function is to serve as a repository of plants. Like butterflies in a cabinet, the effect is more pleasing if they are arranged in some sort of order, but artefacts and edifices are acceptable only if they serve to accommodate the plants. The other extreme is to suggest that a garden is an extension of the house and, as such, is an integral part of the architecture. Plants, where they are used, must not interfere with the architect's line.

Jekyll's approach was based firmly upon a knowledge and understanding of each plant's individuality. She was a highly observant plantswoman, despite her visual handicap, and she based her planting on purely natural principles. Lutyens was designing buildings for an empire. His boulevards in New Delhi are a manifestation of the arrogance of that era. Many of his "landscape" designs are equally selfconscious. For example, at Marsh Court, Hampshire, he made a sunken garden with hundreds of steps, a score or so of small, raised terraced beds, yards of balustrade and a rectangular pond. Nobody, not even Jekyll, would be able to soften this rigid layout with naturalistic planting. And yet the restriction of planting area seems, strangely, to work. There are many good examples of gardens which, although they are not carried to Lutyens' extremes, rely on rigid formal layout within which planting is based upon Jekyll lines. (Sissinghurst Castle in Kent, discussed below, is probably the finest example.) The Jekyll-Lutyens partnership continued for many years, during which more than 100 gardens were planned.

Photography became one of Jekyll's hobbies, and this must have provided some small compensation for her being unable to paint. She certainly took some good pictures, developing and processing the plates herself.[16] Her books continued to appear, and she wrote for various periodicals. All her gardening advice was based on her own practical experience. She was quite used to hard physical work and, although there was no shortage of skilled labour at Munstead Wood, she did a great deal of the gardening herself. Her love of flowers was all-embracing. She had as much affection for the humblest as

VITA SACKVILLE-WEST

A Turbulent Life: a Tranquil Garden

—1892—
Born at Knole, Kent

—1913—
Married Harold Nicolson. Restored garden at Cospoli, Turkey

—1914—
First son born

—1917—
Nigel Nicolson born

—from 1920—
Developed garden at Long Barn, Kent

—1926—
Published The Land

—1929—
Nicolson left diplomatic service

—1930—
Bought Sissinghurst Castle

—1933—
Gave her first garden talk on radio

—1946—
Began regular garden column in The Observer *(ran until 1961)*

—1949—
Elected to National Trust Garden Committee

—1962—
Died

for the grandest. See what she says of London pride:

> *When its pink cloud of bloom is at its best, I always think it the prettiest thing in the garden.*[17]

Of *Gypsophila*:

> *Its delicate masses of bloom are like clouds of flowery mist settled down upon the flower borders.*

And even of Elder:

> *I am very fond of the elder-tree. It is a sociable sort of thing.*

SELF-TAUGHT EXPERTS

Gertrude Jekyll lived to be almost 90. She was still writing a few weeks before she died. To this day, her influence is universal. We are all, to some extent, ruled by her colour schemes, her plant associations, her passion for simple flowers and her dislike of "overdoing" things. Examples of Jekyll-inspired gardens abound everywhere. Vita Sackville-West, although she met Jekyll only once, was probably deeply influenced by her. Her garden at Sissinghurst certainly bears all the Jekyll trademarks, even though the two women were not in direct communication. As Robinson and Jekyll are among the most prominent horticulturists of the day, it would have been unlikely for any gardener in the early part of the 20th century not to have been influenced by them.

Victoria (Vita) Sackville-West

led such a strange life that it is easy to become fascinated with her biography and forget about her plants. She was decidedly upper-crust — her father was Lord Sackville. She was born at Knole, which is a 1,000-acre (400ha) estate on the edge of Sevenoaks, Kent. Her parents were cousins, each of whom had affairs, and they separated. Her mother became mentally ill towards the end of her life.

The house at Knole is enormous, with the parkland running up to lawns and then the lawns running up to the house. Vita's only gardening in her childhood and youth consisted of sowing cress on a wet flannel and growing a few childish vegetables.[18] She married Harold Nicholson, son of Lord Carnock and nephew of the Viceroy of India (Lord Dufferin), at Knole in October 1913. Nicolson was a diplomat, and they moved straight to Turkey where, at Cospoli, they began to develop their first garden. They had very little time there, however, for Nicolson was recalled in 1914.

In 1915 they bought Long Barn, a cottage quite close to Knole. It was a mess. The house was in poor condition and the land was piled with rubble and weeds. They set to work, repairing the house and creating a garden from the chaos outside. At this stage, Sackville-West knew nothing at all about gardening or plants. She wrote in her notebook: "When and how to plant lilac? When wild thyme? Wild sedums? ... What other good rock things, bushy? Good climbing roses?"[18] However, within a few years she had served an energetic apprenticeship and had absorbed enough knowledge to be able to grow most things successfully. She was especially fond of wild flowers and planned, at Long Barn, wild rose and honeysuckle in the hedges and wild cherry in the copse.

At the end of World War I, their private lives were in disarray. Harold had become infected from a homosexual adventure and Vita had an affair with Violet Keppel, daughter of one of Edward VII's mistresses. The two women took

The Nicolsons created the gardens at Sissinghurst from a ruin. This is the so-called cottage garden in the grounds.

a number of holidays together over a three-year period. In the meantime, Violet Keppel married Denys Trefusis but resumed her relationship with Sackville-West. Eventually, the affair fizzled out, but there were other affairs with other women — including Virginia Woolf. Nicolson and Sackville-West patched things up and their marriage, such as it was, continued.

In 1926 Sackville-West published a long poem, *The Land*, which was critically acclaimed and won her the Hawthornden Prize for Literature. She wrote a number of other works of both poetry and prose.

DERELICTION TO DELECTATION

In 1929, when development was threatened near Long Barn, the Nicolsons went house-hunting. Near Sissinghurst, Kent, they looked

Roundel hedges, seen from the tower at
Sissinghurst Castle.

at a wrecked Elizabethan building, a huddle of
broken-down farm buildings and an odd-
looking tower. For this assortment, plus 500
acres (200ha) of land they paid £12,000. By
today's standards this may sound cheap, but it
was not.

The property needed £15,000 spending on
it, and even then accommodation was likely to
be inconvenient. However, the place had distant family connections and, as Nicolson wrote
to Vita, in part:

> *c) It is in Kent. It is a part of Kent we
> like. It is self contained, I could make a
> lake. The boys could ride.*
> *d) We like it.*

From then onwards, the Nicolsons spent
much of their time and energy creating, out of
this delapidated site, one of the finest gardens
in the country. Nicolson was the designer. His
ideas of line and contour were impeccable. His
was the rounded hedge, his the pleached lime
walk where the spring garden was to be. He
had a lot going for him. Sissinghurst Castle was
full of old, mellow brick walls and courtyards, a
dry moat and lots of different levels. By laying
out a crafty network of hedges he was able to
divide the area into different zones, each inde-
pendent, but each clearly leading to the next.

Vita Sackville-West, with ten years' experience at Long Barn behind her, became the
plantswoman. There is a great deal of Jekyll in
her planting, particularly in her use of colour.
Her triumph is the rose garden, where she fos-
tered many varieties of old rose that had gone
so far out of fashion they were in danger of
extinction. She and Nicolson rescued one
unfamiliar rose which was growing in the rub-
bish when they moved in; it was thought by
some to be ' Rose de Maures ', previously considered extinct, and was reintroduced as ' Sissinghurst Castle '.

She used a variety of plants in association
with the roses: irises, dittany and *Alchemilla*.
She planted *Allium albopilosum*, whose lav-
ender-mauve flowers go so well with the blush
shades of the old roses, and whose bold,
globe-shaped seed-heads keep the interest
going after the roses have finished.

She laid out a white garden — very Jekyll! —
with brick paths and centred by a huge
umbrella of *Rosa longicuspis* which is dazzling
in June. To heighten the whiteness she planted
silver-foliaged subjects and covered every inch
of soil with ground-cover — white comfrey and
white *Pulmonaria*.

From 1946 onwards, Sackville-West wrote a
regular column for *The Observer.* By now she
was recognized as a leading figure in horticultural circles. Her readers, although they may
not have aspired to making their own Sissingh-
ursts, were able to identify with her. For
example, on the universal problem of sparrows
attacking primrose flowers, she wrote:

Has any reader of these articles a sovereign remedy against this naughty, wanton, wild destruction? ... This is a real S.O.S. I have quite a collection of uncommon primroses, Jack in the Green, Madame Pompadour, Cloth of Gold, and so on but what is the good of that if the sparrows take them all?

Nobody provided an efficacious cure. Someone even suggested Christian Science.[15]

By 1961 Sissinghurst was open to the public regularly, and was a delight to stroll through at any time of the year. Vita Sackville-West, now nearly 70, was forced to give up her writing because of illness. She died of cancer in 1962. The garden belongs to the National Trust, who maintain it, as faithfully as possible

Modern gardening along naturalistic lines is successful only when a rich variety of interesting plants is grown. The Nicolsons and Gertrude Jekyll enjoyed plenty of choice but, since their passing, we have benefited from an almost endless supply of new hybrids and new introductions of wild plants from all over the world. Throughout gardening history, keen collectors have always played an all-important role. Their work is all the more useful when based on practical trials in their own gardens.

CROCUSES, SNOWDROPS AND TULIP TEAS

There were many important collectors at the beginning of the 20th

E.A. BOWLES

The Last Great Plantsman

—1865—
Born May 14th, Myddelton House, London

—1889—
Visited Italy and collected first plants

—1900—
Elected to Scientific Committee, Royal Horticultural Society (Crocus specialist)

—1908—
Elected to Council of Royal Horticultural Society

—1910—
Discovered Primula bowlesii; made frequent alpine trips over the years

—1914—
Published My Garden in Spring and My Garden in Summer

—1915—
Published My Garden in Autumn and Winter

—1923—
Published A Handbook of Crocus and Colchicum for Gardeners

—1934—
Published A Handbook of Narcissus; technical collaboration with Stearn on anemones and Stern on snowdrops

—1954—
Died (May 7th)

One of Bowles' illustrations, painted in 1917, of a hybrid of *Galanthus plicatus* and *G. elwesii*.

century — men like E.H. Wilson, Reginald Farrer, William Purdom and Frank Kingdon-Ward — but one who was not only a collector but a fine gardener as well was E.A. Bowles. Whenever anyone begins to develop a burning interest in garden plants — something which can happen at almost any age — it is not long before the name "Bowles" crops up. Take any nursery catalogue, visit any public garden or leaf through any book about gardening and, sooner or later, you will find at least one plant named after him. Think of *Viola* 'Bowles Black', *Crocus chrysanthus* 'E.A. Bowles' (see page 80), *Cheiranthus* 'Bowles Mauve' or *Cyclamen hederifolium* 'Bowles Apollo'.

Edward Augustus Bowles was born, lived and died in the same house. He was never short of money and never had to earn a living. Many in his position would have contributed little to society but, not counting his extensive work for charity, Bowles provided horticulture with a superb legacy of plants. His garden writing is steeped with expertise but at the same time entertains. His rugged support of the Royal Horticultural Society over half a century has been too valuable to calculate.

In the early 18th century Sir Hugh Myddelton constructed a "New River" to carry sweet water to the mushrooming population of London. Bowles' father was the last governor of the New River Company, and lived in the handsome and roomy Myddelton House on the riverbank. The company held its board meetings at the house, and young Bowles was befriended by one of the board members, Canon Henry Ellacombe, who was a knowledgeable plantsman and garden writer. Bowles, having decided to take Holy Orders, had read theology at Cambridge and was embarking on a course at theological college when his elder brother died of tuberculosis; his younger sister also caught the disease, and died soon afterwards. Bowles abandoned his training and returned to look after his bereaved parents.[15] Although he never became a priest, he was thereafter to work industriously at many pastoral duties. He ran a night school for boys as well as a Sunday school, and he took on far more than his fair share of parish duties.

He began plant-hunting abroad in his early twenties, becoming especially interested in the genus *Crocus*. By 1901 he was growing 135 species and varieties of crocus in open frames.[15] He became a close friend of Reginald Farrer (practically the father of modern rock gardening) and shared his enthusiasm for alpines. Bowles' rock garden at Myddelton House had to be big enough to house a wide and growing collection of alpine shrubs such as *Daphne col-*

lina (see page 86) as well as tiny plants and bulbs of all sizes from snowdrops to the tall and graceful *Dierama pulcherrimum* (see page 90). As an asthmatic and hay-fever sufferer, Bowles was in the habit of taking off for the mountains at the time of year when pollen counts in London reach their peak.

In 1900, because of his work with crocuses, he was invited to join the Royal Horticultural Society Scientific Committee — quite an honour for someone in his early thirties with no scientific qualifications! He served the society diligently and in 1908 was elected to its Council. He attended meetings for 45 more years. In 1917 the society awarded him the Victoria Medal of Honour.

Bowles' strength was in his ability to identify a plant quickly and accurately. Although, like Gertrude Jekyll, his eyesight was by no means perfect, he had acute powers of observation. As a talented artist he was able to illustrate his own books on crocuses, *Colchicum* and daffodils. His drawings are not only botanically accurate, they are a delight to look at. He was an amateur, untrained botanist, but this was no disadvantage. Indeed, the cold science of botany was a little hard for him to take at times. When working with Sir Frederick Stern on a book about the genus *Galanthus* (snowdrops), he found the technical approach so irksome that his contribution was restricted to a chapter on how to grow them in gardens.

His main contribution to garden literature was the trilogy *My Garden in Spring* (1914), *My Garden in Summer* (1914) and *My Garden in Autumn and Winter* (1915). The works are packed with useful information, but they are also humorous and entertaining. As a reader you are conducted round the garden with Bowles' gentle commentary flowing as you go. His powers of observation show you things you would never have noticed. Take his description of the scent of *Cytisus battandieri*:

Sometimes it reminds me of strawberries, and at others of grapefruit and lemons, or of a fruit salad with a dash of maraschino or kirsch. You can get all these scents from the same bunch of blooms at different times of the day.[15]

The sunken garden at Hampton Court beautifully maintained today, showing modern tulips used as carpet bedding.

Bowles was one of the few men to see the funny side of some plants. He even had what he called a "lunatic asylum", for plants that had gone off their rockers: black pansies, oak-leaved laburnum, evergreen elder and twisted hazel. His garden may not have exhibited the finest landscaping, but it certainly held one of the most exciting collections of plants this century has seen in private hands.

Living on, gardening to a ripe old age, Bowles was popular and respected. The church fêtes at Myddelton must have provided a wonderful excuse for budding plantsmen to browse in this little paradise. Every year on his birthday a "Tulip Tea" was given. The tulips were grown in a series of raised beds along the New River, each bed underplanted with massed forget-me-nots. In his late eighties, although still active, he began to have heart trouble. The frail old gentleman managed to sit on a Royal Horticultural Society Floral Committee meeting for the last time in April 1954. He died the following month, a few days before his 89th birthday.

CONTINUING
THE GARDEN TRADITION

*I*N the short space available we have but glanced at a handful of the great number of men and women whose talents have helped to bring a wealth of colour, scent, grandeur, charm and interest to our gardens. Many contributors are uncelebrated because there are no records. Not every gardener had the ability to write, and very few were lucky enough to have details of their achievements set down on paper.

The early Christian monks carefully preserved valuable herbs long after the Romans had faded out of history. There were many more 16th-century explorers than mentioned in this text, all risking their lives in the search for knowledge.

There were the florists, working people whose hobbies included improving flowers by breeding and selection. The Paisley weavers worked on pinks. Lancashire mill workers had auriculas. Many modern hybrids are descended from florists' varieties which, in the 19th century, artisans might trade for more than a week's wages. Tulips were bred in this way, and ranunculus, polyanthus and pansies.

There were the rose lovers, breeding, con-serving and promoting roses through the centuries from the time of the Crusaders to the present day — rosarians like the Rev. J.H. Pemberton, who produced the hybrid musks 'Penelope' and 'Buff Beauty' in the 1920s. There were gardeners like Ernest Markham, employed by Robinson, with whom we associate fine cultivars of *Clematis* — 'Ernest Markham' and 'Markham's Pink'. Or the Rev. William Wilks who produced his strain of poppies and who, under Sir Trevor Lawrence, helped to reconstruct the Royal Horticultural Society after its near collapse in 1888.

In these times of vanishing species it is all the more important that great botanic gardens of today continue the work of conserving and collecting: Huntington, California, with its Myrtle, Cambridge with its Geraniums and Fairchild, Florida, with its palms and cycads are three examples.

Finally, there are the silent workers of horticulture — ordinary people who, year after year, tend their gardens, support local flower shows, exchange plants with one another and, through mutual enthusiasm and energy, help to keep gardens interesting and enjoyable.

A-Z
DIRECTORY OF PLANTS

*I*N the first section of this book, devoted to the great horticulturists and botanists, we have seen evidence of a developing awareness of the central importance of plants to Man's physical and emotional well-being (as balm both for the sick body and the restless soul), and that this awareness prompted a strong urge for enlightenment. In different centuries the desire for knowledge has taken various forms: in the early days it was the medicinal properties of plants which absorbed the energies of educated minds whereas, by the late 18th century, aesthetic considerations coupled with the search for successful cultivation techniques, rose in the scale of priorities. This last is demonstrated by the single-minded pursuit of excellence in a narrow field as characterized by the growers of "florist's flowers", such as the auricula and the pansy. By the 1870s, the emphasis had shifted towards finding good garden-worthy plants in preference to those which were medicinally important or interesting only to the "florist". Plant hunters went out from Europe to search all over the world for plants to adorn gardens and glasshouses. Plant-collecting reached its peak in the early years of the 20th century. The wealth of plant material available today is, in part, a tribute to the determination and courage of plant collectors over the last 150 years, and the immense lengths to which rational people will go to in pursuit of a praiseworthy goal.

Many of the species described in the A-Z Directory, such as the pomegranate and the opium poppy, have been cultivated for thousands of years. For those who believe that the enjoyment of gardening lies not solely (or even principally) in the practical cultivation of plants, but also in the aggregation of information which concerns them, the origins and history of these plants is undoubtedly fascinating. Any stroll around the garden, especially if shared with friends, becomes all the more precious and rewarding.

The fifty plants in the A-Z Directory all have some connection with the horticulturists, plant collectors and botanists featured in the first section of the book: they have either been discovered by, written about, or grown by these enlightened men and women. Each entry contains information about how the plant reached us (where applicable), whom we can thank for it, its culinary or medicinal uses, and any stories or legends which are connected with it. A short description of the plant is given, and is written with the general-interest reader as well as the gardener or botanist in mind. Interesting hybrids or forms grown in gardens today are also mentioned here. Practical advice for the gardener is given in the form of a short account of the best means of cultivating the plant, its preferences in the way of soil and aspect, as well as the best means of propagating it. Lastly, there is a quick reference section on the main features of the plant: its size, flowering time, flower-colour and scent.

ACANTHUS MOLLIS Bear's Breeches   *54*

ACER DAVIDII David's Maple   *56*

AGAPANTHUS AFRICANUS (A. UMBELLATUS) African Lily   *58*

ANEMONE ROBINSONIANA Robinson's Windflower   *60*

AQUILEGIA CANADENSIS Columbine; Granny's Bonnet   *62*

ARBUTUS UNEDO Strawberry Tree   *64*

ASTER NOVAE-ANGLIAE   *66*

BUDDLEIA DAVIDII David's Butterfly Bush   *68*

CALLISTEMON SPECIES Australian Bottle-brush Plant   *70*

CAMELLIA JAPONICA Common Camellia   *72*

CHAENOMELES SPECIOSA   *74*

CLEMATIS ALPINA (ATRAGENE ALPINA)   *76*

CROCUS CHRYSANTHUS   *78*

CYCLAMEN PERSICUM Sowbread   *80*

DAHLIA COCCINEA   *82*

DAPHNE COLLINA   *84*

DIANTHUS PLUMARIUS Garden Pink   *86*

DIERAMA PULCHERRIMUM Wand Flower; Angel's Fishing Rods   *88*

DIGITALIS PURPUREA Foxglove; Witch's Thimble; Finger-flower   *90*

FRITILLARIA IMPERIALIS Crown Imperial   *92*

GALANTHUS NIVALIS Common Snowdrop; Milk Flower   *94*

GARRYA ELLIPTICA The Quinine Bush; The Silk Tassel Bush; Fever Bush   *96*

HELLEBORUS NIGER Christmas Rose   *98*

IRIS GERMANICA Purple Flag; London Flag   *100*

KNIPHOFIA GALPINII Galpin's Red Hot Poker: Torch Lily   *102*

LILIUM CANDIDUM Madonna Lily   *104*

MAGNOLIA×SOULANGIANA   *106*

MECONOPSIS GRANDIS   *108*

MYOSOTIS ALPESTRIS   *110*

MYRTUS COMMUNIS   *112*

NARCISSUS PSEUDONARCISSUS Lent Lily; Wild Daffodil   *114*

NICOTIANA ALATA (N. AFFINIS) Tobacco Plant   *116*

PAEONIA OFFICINALIS   *118*

PAPAVER SOMNIFERUM Opium Poppy   *120*

PHILADELPHUS CORONARIUS Mock Orange   *122*

PLATANUS ORIENTALIS Oriental Plane; Oriental Sycamore   *124*

PLATYCODON GRANDIFLORUS Balloon Flower; Chinese Bellflower   *126*

PRIMULA AURICULA HYBRIDS Dusty Miller; Bear's Ears; Mountain Cowslip   *128*

PRIMULA FLORINDAE Giant Cowslip   *130*

PRUNUS 'TAI-HAKU' Great White Cherry   *132*

PUNICA GRANATUM Pomegranate   *134*

RHODODENDRON ARBOREUM The Tree Rhododendron   *136*

'ROSA MUNDI' (R. GALLICA 'VERSICOLOR')   *138*

SALVIA PATENS   *140*

SCHIZOSTYLIS COCCINEA Kaffir Lily; Crimson Flag   *142*

SORBUS 'JOSEPH ROCK' Joseph Rock's Rowan   *144*

SYRINGA PERSICA Persian Lilac   *146*

TULIPA CLUSIANA The Lady Tulip   *148*

VIBURNUM TINUS Laurustinus; Wild Bay   *150*

VIOLA TRICOLOR HYBRIDS Garden Pansies   *152*

ACANTHUS MOLLIS

Bear's Breeches

*T*HIS plant, which is a native of Italy and southern Europe, was known in Britain long before its introduction there in 1548. It soon became popular because of its supposed efficacy as a soother of burns, gout and stomach upsets. John Evelyn wrote that it was a "mollifying herbe" for "members out of yointe".

This is the plant whose leaves are depicted, in a stylized way, at the top of Corinthian columns. The story goes that the architect Callimachus (late 5th century BC), who was at the time building a temple in Corinth, saw a plant of it growing under a basket on which had been placed a tile. Apparently the leaves had grown through the basket and been turned back by the tile. A picturesque story, but little credence need be given to it.

In 1551, William Turner noted in his *New Herbal* that

> *this herbe groweth plentifully in my Lordes* [*Somerset*] *garden at Sion* [*Syon Park*]. *I never saw it grow wilde as yet* [*It was not found wild in England until 1820*] ... *They that will have anye more of the description of Branke Ursine* [*bear's claw*] *let them rede the description of Dioscorides ... which I do nowe pass over because I know that the herbe is so perfitelye knowen in all countries.*

It may not be "so perfitelye knowen" to all as, until recently, it was not widely grown. The fashion for planting striking "architectural" plants, however, has brought it once more to public notice and acclaim. William Robinson, for example, called it "long neglected" in his *The English Flower Garden* of 1883. The plant has large, mid-green, shiny, pendulous, deeply indented leaves. The flowers, on long, upright spikes, are white with purple hoods (bracts). It is a most stately and imposing plant. These days, *A. mollis* is usually represented in gardens by its stronger-growing variety, *A. mollis latifolius*. Also widely grown is the attractive. *A. spinosus* (shown opposite).

CULTIVATION

This plant is not difficult to grow, provided that one leaves it to its own devices and resists all temptation to move it around. It is happiest left in peace in either a sunny or a partially shaded place in a deep, well drained soil. The soil need not be rich, but it should be deep because the roots will venture down a long way. A little winter protection in the way of peat or straw is advisable in cold districts. *A. mollis* can be propagated quite easily by sowing seed in the autumn or spring in pans, which are put in a cold frame; the seedlings can be planted out in rows and transferred to their permanent positions in two years' time. The quicker method is to take root cuttings in winter, or to divide the roots and replant them in the dormant season.

FEATURES

Height: 3-5 feet (90-150cm)
Flowering period: July and August
Colour: White and purple flowers; glossy green leaves

1 Petal
2 Style
3 Stamen
4 Sepal

ACER DAVIDII

David's Maple

*A*CER DAVIDII comes from China, where it has a wide distribution, and has been grown in Britain only for the last century or so. The plant collector Maries, who was employed by the famous Veitch nursery, sent seed of it home from the province of Hupeh in 1879. Some years later E.H. Wilson, collecting in the same region, also found it and sent it back. It was found later by George Forrest and Frank Kingdon-Ward in Yunnan, southern China; these southern plants do not look the same as the Hupeh introductions, having larger leaves and reddy-purple young stems.

The main features of this small but handsome deciduous tree is the very obvious striping on the bark. This has given *A. davidii* (along with some similar maples, such as *A. capillipes, A. rufinerve* and *A. pennsylvanicum*) the name of 'Snakebark Maple'. The mature shoots are grey-green, striped with white. The leaves are dark green, up to 7 inches (18cm) long, oval in shape, but not lobed. They have a reddish tinge to them when breaking in the spring, and the autumn colour, usually yellow and red, is good enough to make this a highly desirable small tree for any garden.

Père Armand David, after whom this plant was named, was a French missionary who, like many missionaries, was a good naturalist. He eventually gave up his mission and collected plants and animals for the Museum of Natural History in Paris instead. Both the 'Handkerchief tree', *Davidia involucrata*, and a species of small deer, Père David's deer, were named in his honour. He was the first European to send back information about the Giant Panda.

CULTIVATION

A. davidii, like many a Chinese species, is tolerant of an alkaline soil. It does, however, benefit from an unexposed position where its bark is not cracked or its leaves burned by sunshine after frosty nights. The beauty of its autumn colour demands it be placed in a sheltered place and one where wind will not dislodge the leaves prematurely. The bark is striped best if the tree is grown in a partly shaded place. Although generally healthy, *A. davidii* can be attacked by aphids in spring. Seed should be sown in a cold frame in autumn so that the germination inhibitors in the seed can break down during the winter. Once that has happened, the seed will germinate quite freely. Most maples do not root very well from cuttings.

FEATURES

Height: 30-50 feet (9-15m)
Flowering period: May
Colour: Yellow flowers; green and yellow fruits

1 Pericarp
2 Seed coat
3 Cotyledon

1 ——————————
3 ——————————
                                    —— 2

AGAPANTHUS AFRICANUS (A. UMBELLATUS)

African Lily

*T*HIS was introduced to Britain from the Cape Province of South Africa as early as 1629. The first illustration of it, dated 1692, is by Plunkenet; the model for his illustration was flowering at Hampton Court, and so may well have been imported by the nursery firm of London and Wise. The nomenclature is a little uncertain, having changed from *A. africanus* to *A. umbellatus* and back again.

The African Lily is a half-hardy perennial with fleshy roots, and flowers (up to 30 in an umbel) which are borne on the end of long succulent stems above a clump of strap-shaped evergreen leaves. The flowers are deep violet-blue and like an open bell in shape. This plant is reliably hardy only in the very south and west of Britain, which is why during this century hybrids have been raised that are capable of being more widely grown. The most famous of these are the 'Headbourne Hybrids' raised by the Hon. Lewis Palmer in the 1950s and 1960s at his garden at Headbourne Worthy, near Winchester. These have violet-blue or light blue flowers.

CULTIVATION

Agapanthus are really only just hardy in the British climate, and benefit from being protected in, say, a sunny wall border, with a blanket of peat or ashes put over them in autumn. They are also most suitable for growing in pots or tubs in the garden provided they are brought into the cool greenhouse to be overwintered. They like a fertile, deep, but well drained soil, and their crowns should be planted about 2 inches (5cm) below ground level in April. Regular feeding with a liquid fertilizer in the growing season will encourage them to flower well. While flowering, they should be watered well in dry periods.

They are best and most easily propagated by division in late spring as they readily make offsets, which can be detached. Seed can be sown in heat in spring, but the seedlings will take from three to five years to flower.

FEATURES

Height: 2-2½ feet (60-76cm)
Flowering period: July to September
Colour: Deep blue-violet

1 Petal
2 Stamen
3 Style
4 Ovary

ANEMONE ROBINSONIANA

Robinson's Windflower

ACCORDING to William Robinson (never a man to hide his light under a bushel) in *The Wild Garden* (1870):

> The most beautiful form of our wood Anemone [Anemone nemerosa] *which has come into the garden in our day is the large sky-blue form. I first saw it as a small tuft at Oxford [Oxford Botanic Garden] and grew it in London where it was often seen with me in bloom by Mr. Boswell Syme, author of the Third Edition of Sowerby [James Sowerby's* English Botany, *first published in 36 volumes, from 1790 to 1814] ... and we were often struck with its singular charm about noon on bright days. There is reason to believe that there is both in England and Ireland a large and handsome form of the wood Anemone — distinct from the common white of our woods and shaws [small woods] in spring, and that my blue Anemone is a variety of this. It is not the same as the blue form wild in parts of North Wales and elsewhere in Britain, this being more fragile looking and not so light a blue.*

He thought it useful for the rock-garden, for the edges of borders, for the wild garden beneath shrubs and for naturalizing in grass.

E.A. Bowles believed that *A. robinsoniana* arose not from Ireland but from Norway, via Ireland. Being something of a botanical sleuth he was probably right. He could never

> *forgive it for closing so readily on dull days and towards evening in a rather sulky, short-tempered way, and then displaying what I can only call a cotton back, a poor greyish-tinted outside not much better than a dirty white. When open its soft, glowing, rosy-lilac flowers are certainly very lovely and a large patch of it glistening with April raindrops in its leaves, but its flowers open to the sunshine, makes one want to cease worrying about weeds and just enjoy the Spring scents and flowers.* [My Garden in Spring, 1914.]

However, Bowles knew a form of *A. robinsoniana* called *A.r. Allenii* which was everything that he could desire, as it was flushed with rosy-purple on the outside.

A. robinsoniana has flowers which are more than an inch (2.5cm) across and lavender-blue; the petals have edges which are slightly undulated. It has leaves which are cut into three segments and very deeply toothed. It flowers in March and is a glorious sight when grown in quantity as a carpet under deciduous trees.

CULTIVATION

Closely connected as it is to *A. nemerosa*, *A. robinsoniana* revels in the conditions of moist woodland which suit our native Wood anemone. However, as Robinson noted, its uses in the garden are manifold, as it is obviously a tolerant and well disposed plant. To propagate it, the rhizomatous rootstock can be divided when the foliage has died down completely in summer, and planted 2 inches (5cm) deep. An annual mulch of leaf-mould is beneficial.

FEATURES

Height: 6-8 inches (15-20cm)
Flowering period: March and April
Colour: Lavender blue

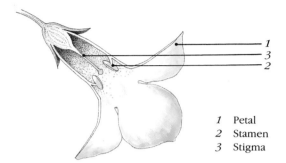

1 Petal
2 Stamen
3 Stigma

AQUILEGIA CANADENSIS

Columbine; Granny's Bonnet

*A*QUILEGIA CANADENSIS was the first aquilegia to be introduced to Britain. It comes from northern North America, and its introduction is one of the results of the fruitful plant-hunting trip which John Tradescant the Younger made to Virginia in 1637. It differs from our native Columbine in having yellow flowers with red spurs to the petals. Linnaeus was the first person to discover that these spurs contained the plant's nectaries. The flowers are up to 1½ inches (4cm) across and half an inch (1cm) long, with longer spurs, and come out in May and June. The leaves are in three parts and dark green. Like all other columbines, *A. canadensis* hybridizes promiscuously with others in the garden. There is a dwarf form, *A.c.* var. *nana*, as well as a paler yellow version called *flavescens*.

CULTIVATION

A. canadensis enjoys a moist, but not waterlogged, soil in sun or a little shade. Like all columbines, it will hybridize with others easily, so the stems need to be cut down after flowering to prevent that. It is best propagated by sowing the seed when it ripens in late summer, or in the following March, in seed trays, which can be put in a cold frame; the seedlings can then be pricked out when large enough to handle. Division is possible in the dormant season if you wish to avoid any hybridization. Aquilegias are occasionally bothered by aphids but are otherwise reasonably trouble-free.

FEATURES

Height: 12-24 inches (30-60cm)
Flowering period: May and June
Colour: Yellow with red spurs

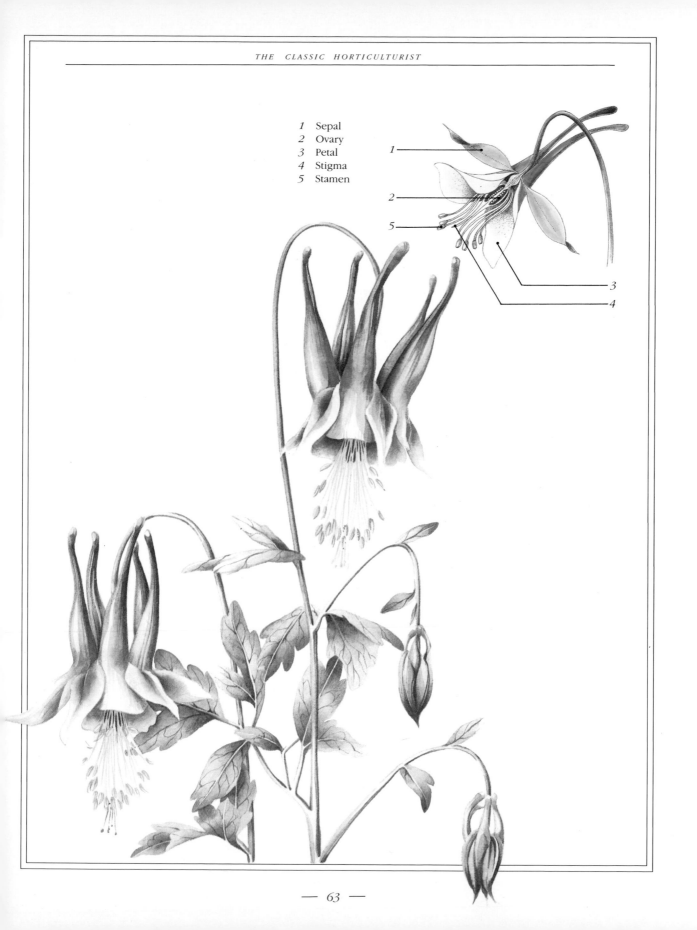

1   Sepal
2   Ovary
3   Petal
4   Stigma
5   Stamen

ARBUTUS UNEDO

Strawberry Tree

*A*RBUTUS UNEDO has a most interesting history. It has been known in Britain at least since the 16th century. It is a native of the Mediterranean region and Asia Minor (where its berries are used to make preserves and alcoholic drinks, and the bark, leaves and fruit are used for tanning) as well as of the southwest of Ireland, around the Lakes of Killarney. There is a story to explain why this has come about: the tree appeared in Ireland as a miracle to remind the monk Bresal of the time he had spent in Spain teaching the Spanish monks Irish choral music. Appealing as this explanation is, it is more likely that the arbutus grew all over Europe just after the last retreat of the glaciers (about 10,000 years ago), when the climate was warmer than it is now, but that, as temperatures cooled in northern Europe, it survived in only a few places.

This arbutus was described by Turner in 1548, but it may have been in Britain before that time, for it is thought that the Romans might have brought it with them. In 1759 Philip Miller was listing not only a red-flowered kind but also double-flowered varieties. Parkinson remarked that the colour of the fruits was like a "pallide clarret wine".

The plant's species name, *unedo*, means "one I eat"; this refers to the fruits of the plant. Although these are edible, they are not sufficiently nice to tempt anyone to eat more than one.

A. unedo fruits only in mild districts — probably because it flowers between October and December, a time when there are not many insects active to effect pollination. It fruits at the same time as it flowers, which makes it (especially considering the time of year) a most valuable ornamental shrub or small tree. The flowers are usually white or pink and hang in short pendant panicles. The fruits are bright orange-red and round, up to three-quarters of an inch (2cm) in diameter, and their colour is the reason why the plant is called the Strawberry Tree. This tree, unlike the Madrona, which has a marvellous smooth cinnamon-coloured bark that peels off in large flakes, has a rough bark and elliptical or oval leaves, which taper at both ends. The leaves are smooth and dark green.

CULTIVATION

In general terms, *Arbutus* are not happy on alkaline soils, but *A. unedo*, which is found wild on limey soil in Yugoslavia, is an exception. All *Arbutus* species require a sunny, sheltered position in a deep, well drained soil and some winter protection when they are small and tender. They are usually propagated by semi-hardwood heel cuttings, put in a heated propagating case, or by ripe seed sown in March and placed in a cold frame. The young plants should be grown on in the frame until large enough to be transplanted in spring. *A. unedo* does not generally need pruning.

FEATURES

Height: Up to 30 feet (9m)
Flowering period: October to December
Colour: White or pink; fruits orange-red

1   Stigma
2   Sepal
3   Petal

ASTER NOVAE-ANGLIAE

*A*STER NOVAE-ANGLIAE was introduced to Britain in 1710 — i.e., quite late, when we remember that *A. amellus* has been growing in our gardens since the end of the 16th century. *A. novae-angliae* comes, as its name suggests, from New England. Asters were originally called Starworts, but the name changed gradually to Michaelmas Daisies some time after 1752. This was when the new Gregorian calendar was introduced, with the result that Michaelmas (September 29th) now fell 11 days earlier and began to coincide with the time when these asters were flowering. Perennial asters were not very popular until William Robinson championed their cause — and those of many other hardy herbaceous perennials — in reaction to the obsessive English interest in half-hardy annuals.

A. novae-angliae has drab green lanceolate leaves and clusters of flowers up to 2 inches (5cm) in diameter; these appear in September and early October. It hybridizes less promiscuously with its relations than do other asters, so there are fewer modern cultivars: the best include 'Barr's Pink', 'Harrington's Pink' and 'September Ruby'. Its great advantage is that it does not suffer from aster wilt; nor is it so prone to mildew as some other species.

CULTIVATION

Asters like a well drained but not droughty soil which is well dug with organic material. They prefer sun, although they will stand a little shade. They usually need staking, as well as mulching in spring to prevent the soil drying out. *A. novae-angliae* has not the need for annual division which makes *A. novi-belgii* rather a nuisance to grow, but it should be divided every three years in the dormant season. When the plant is being divided, the middle piece should be thrown away and only the younger outside buds replanted. Slugs like them, as do caterpillars, but powdery mildew is rarely a problem.

FEATURES

Height: 4-6 feet (120-180cm)
Flowering period: September to October
Colour: Pink or violet-purple

1   Stamen
2   Petal
3   Sepal

BUDDLEIA DAVIDII

David's Butterfly Bush

*E*VERYONE, even if they never garden, knows the Butterfly bush — although perhaps not by name. A native of China, it has happily established itself in British gardens and moved out to colonize railway sidings and wasteland. With its long tube-like flowers in panicles, it is much loved by butterflies and planted widely for that reason alone. But it is also a very handsome and hardy deciduous shrub.

It was first introduced to Britain by Dr Augustine Henry, an Irish doctor who joined the Imperial Customs Service in China and lived for several years at Ichang. He named it after Père David, who had first found it there some years before. However, a better form was raised, by the French nursery, Vilmorin, from seed sent in 1893 by another missionary, Père Soulié. Père Soulié's fate is an instructive example of how dangerous plant-hunting can be: he was killed by bandits in 1905.

B. davidii makes a strong-growing bush, with grey-green lanceolate leaves and slightly arching panicles of scented lilac flowers, orange at the mouth, which come out between July and October. These panicles can be as long as 20 inches (50cm) on vigorous plants. Numerous good forms have been raised, the best probably being "Black Knight", "Royal Red" and "White Profusion". Of the other Buddleia species, *B. alternifolia* has long arching sprays of lavender-blue flowers in June, and *B. globosa* has round balls of orange-yellow flowers in early summer.

B. fallowiana is grown against a wall because it is not very hardy.

The genus is named after the Reverend Adam Buddle, Vicar of Farnbridge, Essex, and a keen amateur botanist. The connection between men of the cloth and botanical discoveries has been a constant theme in British history. It would be too facile to suggest that they did not busy themselves enough with their parishes; rather they found God's creation endlessly fascinating.

CULTIVATION

Buddleia davidii could not be easier to cultivate, although it takes a little care to make it give of its best. A good fertile soil in a sunny position suits it the best. It flowers on the growths made in the current season, so it should be pruned back to within two buds of the old wood in late February or early March. Cuttings strike readily if taken from short side-shoots, with a heel, in late summer and put in a cold frame. Alternatively they can be taken in October and put in a row outside in a sheltered place. Seed can be sown, but those varieties named will not come true from seed.

FEATURES

Height: 9-12 feet (2.7-3.7m)
Flowering period: July to September or October
Colour: Mauve
Scent: Very sweet

1 Sepal
2 Ovary
3 Stigma
4 Stamen

CALLISTEMON SPECIES

Australian Bottle-brush Plant

IN 1768, Joseph Banks set out with Captain Cook on the *Endeavour* to Tahiti and then on to find out more about the southern land which Tasman had glimpsed so tantalizingly in 1642. After landing in New Zealand, they eventually found Australia, dropping anchor at Botany Bay, on the south coast, in April 1770. With his companion, Dr Daniel Solander, Banks did much good botanizing there before the *Endeavour* sailed north to the Barrier Reef. Among the specimens pressed were species of *Callistemon*. However, *C. citrinus*, *C. salignus* and *C. linearis* were not introduced to Britain as living plants by him but by two men called George Austin and James Smith, whom he sent out from the Royal Botanic Gardens at Kew in 1788. All three "bottle-brush" plants are native to the coastal region of southern Australia, although they differ a little in geographical range.

be grown in very large pots or in the border in a greenhouse or conservatory where the minimum temperature is 7°C (45°F). Callistemons require a light position if they are to flower freely, and need ample water during the summer months. Those grown in pots will need regular feeding while in active growth. Propagation is by cuttings or seed. Cuttings strike most readily if they are taken in late summer (with a heel) and put in a propagating case which can provide bottom heat. Seeds can be sown in March in heat, and pricked off into small pots. The young plants must be overwintered in a frost-free greenhouse and grown on for another year in 5-inch (13cm) pots before they are ready to face the open garden. Callistemons can flourish without any pruning, and their leathery leaves are generally proof against pest and disease problems.

CULTIVATION

The hardiest callistemon, *C. linearis*, can be planted in a sunny border against a warm wall, but in all but the southernmost parts of Britain it should be protected in some way, during the winter. The two other commonest species, *C. salignus* and *C. citrinus*, should

FEATURES

Height: *C. salignus* occasionally up to 30 feet (9m); *C. linearis* up to 5 feet (1.5m); *C. citrinus* up to 15 feet (4.6m).

Flowering period: July

Colour: Crimson or bright red; *C. salignus* is pale yellow

1 Petal
2 Ovary
3 Stigma
4 Stamen

CAMELLIA JAPONICA

Common Camellia

CAMELLIA JAPONICA is a native of Japan, although it has also been cultivated in China for centuries. In 1738, Lord Petre acquired from the East two plants, one single white, the other single red. Grown in a hothouse, these soon died. Fortunately, Petre's gardener had propagated one of them and grew it successfully in a coolhouse in the nursery he opened in the Mile End Road in 1740. Double forms arrived from Asia in later years.

By the 1840s, *C. japonica* was very popular and modish in Europe. "La Dame aux Camellias", Marie Duplessis, who died in 1852, aged 28, always wore camellias, but although she was the most famous person to do so she was only one among many. However, by the 1880s, the flower was losing favour; it was too formal and had no scent.

Until the 20th century, all camellias had to be protected in the winter and grown under glass, or so it was thought. However, people gradually realized that this camellia was in fact hardy, and so they began to plant it outdoors, in woodland settings. In 1930, a camellia called *C. saluenensis*, which had been sent from Yunnan, flowered in the garden at Caerhays in Cornwall for the first time. J.C. Williams, who lived there, crossed it with *C. japonica* and achieved a strain called *C. × williamsii*. One of the best of these is 'J.C. Williams'. Its great advantage over the ordinary camellia is that it drops its dead flower-heads naturally.

C. japonica has leaves which are very dark green and glossy, and oval with shallow toothing. The flowers, which are often semi-double, can be pink, white or red, and are up to 5 inches (13cm) across. Outdoors it flowers between April and June.

CULTIVATION

The main disincentives to growing camellias are their preference for lime-free soil and the fact that they flower so early in the year that their flowers are always in danger from frost damage. Therefore, if the garden soil is alkaline, camellias require to be grown in made-up soil (or, better still, a large tub or pot); and, whatever the soil, the plants should be placed in a sufficiently sheltered place, either against a warm west-facing wall or among trees and other shrubs in woodland, so that the frost has no opportunity to transform their glorious flowers into brown, disintegrating rags overnight. They like a deep, organically enriched soil which will not easily dry out but, at the same time, never becomes waterlogged. Also, they like a cool root run, which is why growing them against south-facing walls is not advisable. They are best planted in early autumn or late spring. They tend to hang on to their dead and dying flower-heads, so these should be removed for the appearance of the plant after the flowers are over.

Camellias can be grown in a border in a cool greenhouse or conservatory. Alternatively, they can be grown in large tubs in which has been put a mixture of 4 parts acid loam, 2 parts peat, and 1 part horticultural sand, together with a sprinkling of bonemeal. These tubs stay outside for the summer and then can be brought into the coolhouse in October, where they should be kept until after flowering. The tubs should not be kept well watered or the flower-buds will drop off. If the leaves begin to yellow or develop spots, apply iron chelates. Propagate by cuttings in the same way as for *Callistemon* (see page 70) or, if the

1 Petal
2 Stigma
3 Stamen
4 Ovary

1 2 3 4

plant has branches near the ground, by layering in early autumn. Layers will root in a year or 18 months. *C. japonica* requires little or no pruning, unless there are inconvenient shoots on wall specimens; these can be cut out in spring. Birds can damage the flower-buds, but the worst enemy is frost; it makes the buds drop and the leaves distort. In the greenhouse or conservatory, camellias are attacked by aphids, mealy bug and scale insects.

FEATURES

Height: Up to 30 feet (9m) in very good conditions, otherwise about 12-15 feet (3.7-4.6m)
Flowering period: April to June outdoors; February to May under glass
Colour: White, pink or red

CHAENOMELES SPECIOSA

CHAENOMELES SPECIOSA is one of many good plants introduced to Britain by Sir Joseph Banks of Kew — or, in this case, sent home to Banks from China by James Main in 1796. Within 40 years it had become a popular plant in British gardens; this is not surprising, considering that it flowers early in the year before the leaves have properly expanded, and that those flowers are semi-double, rosy red in colour, and very showy. It has been called both *Cydonia* and *Pyrus* in the past but has now settled down as *Chaenomeles*. For a long time it was given the common name "japonica", a fact which served to confuse the issue still further. The generic name comes from the Greek words *chaino*, meaning "I gape", and *meles*, meaning "apple", although it is hard to see why or how it received this name.

It is a deciduous shrub which, if left alone, will make a broad shape. It has spiny branches and the leaves are oval and quite glossy above. The flowers, which can be as much as 1¾ inches (4.5cm) across, are produced in clusters on last year's "wood". If the plant is grown against a wall, it is quite possible to have flowers in December, even before Christmas, in mild districts. But the usual time for flowering is February or March, continuing sometimes until the end of May. It occasionally flowers again in the autumn, although not as abundantly. The fruit, which has no stalk, has small dots all over it and smells like the quince, to which the plant is plainly closely related, but it remains green and is not as decorative as the golden fruit of *C. japonica*.

There are some very good varieties of *C. speciosa*. Among the best are 'Moerloosii', which has large white flowers with a pink flush, and 'Nivalis', which has pure white flowers. *C. speciosa* is a parent of many hybrids which have arisen between it and the later-flowering *C. japonica*. These are called, collectively, *C. × superba*, and include varieties like 'Knap Hill Scarlet' and 'Crimson and Gold'.

CULTIVATION

Chaenomeles are easy to grow, requiring only a sunny place in any ordinary garden soil to flourish. If the soil is very alkaline, a dressing of peat each year will probably prevent yellowing of the leaves. Birds can attack the flowers; if this happens, some form of netting should be laid over them as protection. *Chaenomeles* require no pruning if grown as specimen bushes; if trained against a wall, the flowered shoots should be cut back to three buds from the base in early summer. Cuttings of semi-hard wood can be taken in late summer and struck in a propagating case which has bottom heat. Long shoots can be layered.

FEATURES

Height: 6 feet (1.8m); 10 feet (3m) against a wall

Flowering period: January to March if grown against a wall; later in the open

Colour: White, pink, red or crimson, depending on variety

Scent: The fruits smell like true quinces

1 Petal
2 Stigma
3 Stamen
4 Ovary

CLEMATIS ALPINA

(ATRAGENE ALPINA)

CLEMATIS ALPINA is native to Central and Northern Europe as well as Northern Asia. It was introduced into Britain in 1792. Very occasionally, the synonym *Atragene alpina* is used because *C. alpina* belongs to the Atragene group — those clematis with flowers which have, between the sepals and stamens, a ring of petal-like structures called petaloid staminodes. Clematis do not have petals — or, at least, not obvious ones: instead, the flowers are made up of four oval sepals. The staminodes of flowers of the Atragene group can make them look double.

C. alpina is a deciduous climber, with leaves up to 6 inches (15cm) long; these have nine leaflets which are dark green, narrowly oval, and deeply toothed. The flowers, which are violet-blue, are borne in April and May on long stalks and they nod. The four sepals are 1½ inches (4cm) long, the grey staminodes much less. This is not a strong-growing clematis, in comparison with many other species, so it is seen to its best advantage scrambling through a shrub or over a tree stump. There are several good forms of it in cultivation, particularly ' Frances Rivis '.

CULTIVATION

Clematis are twining plants, which means that they naturally clamber over trees and shrubs — hence their need to have their roots in shade. The traditional way of ensuring this is to plant low-growing evergreen shrubs around their bases, especially to the south of them, or to place flat stones or tiles on the ground around the bases. All *Clematis* like a deep, fertile soil and are, contrary to popular belief, not fussy whether it contains lime or not. *Clematis* resent root disturbance, and so they are usually sold in pots which can simply be dropped into a suitable hole in the ground in clement weather between September and April. *C. alpina*, which flowers in April and May on shoots made the year before, needs pruning only when grown in a place where its natural growth may prove a nuisance. If this is the case, prune out the flowered shoots after flowering. All *Clematis* appreciate a mulch in spring of well rotted compost or peat.

Propagation of clematis is achieved most easily by so-called "serpentine" layering. This is the technique whereby a long shoot is laid along the soil in March, and pegged down at several places near the nodes. These layers root in about a year. Cuttings can be taken in July, but require bottom heat to achieve a good striking rate. The rooted cuttings should be overwintered in pots in a frost-free place, the pots plunged in the ground the following spring, and not planted up until the autumn after that.

FEATURES

Height: Up to 8 feet (2.4m)
Flowering period: April and May
Colour: Violet-blue

1 Sepal
2 Ovary
3 Stamen
4 Petal

CROCUS CHRYSANTHUS

THE name of *Crocus chrysanthus* is inextricably linked with that of the greatest gardener-botanist of this century, E.A. Bowles. It is a plant of Greece and Asia Minor, and in the wild is very varied in form. As Bowles put it, in *A Handbook of Crocus and Colchicum for Gardeners* (1924):

> *It is the most variable species known as regards colour, the ground colour being sulphur yellow, orange, white or lilac, while the outer segments show every imaginable degree of freckling, suffusion and feathering of chocolate, brown, grey or purple, or they may be self-coloured.*

This variation gave him ample opportunity and incentive to study this group, to name many of the better forms and cross them with other *Crocus* species. The best he named after birds — for example, 'Yellow Hammer', 'Golden Pheasant' and 'Siskin'. Unfortunately, only 'Snowbunting' is still commercially available. The largest and best, however, is a butter-coloured form, named *C.c. pallidus* 'E.A. Bowles' by the famous Dutch bulb firm van Tubergen, and still widely grown.

Despite its variability, *C. chrysanthus* has some consistent characteristics. The true *C. chrysanthus* always has black barbs at the bottom of the anthers; the throat is always golden; and there are always markings of some kind on the outside of the petals. This last feature adds greatly to the plants' attractiveness, bearing in mind that crocuses do not open their petals unless the sun shines, and the sun shines rarely in the average British winter. For these are some of the earliest crocuses

to flower, usually well up by the middle of February, and lasting into March. Most *C. chrysanthus* grow no more than 3 inches (7.5cm) tall, although 'E.A. Bowles' has a bigger flower than most. The most widely grown forms are 'Ladykiller' (purple-blue, and white within), 'Zwanenburg Bronze' (feathered with bronze, overlaying yellow), 'E.A. Bowles', and 'Blue Pearl' (blue within, almost white outside).

CULTIVATION

The various forms of *C. chrysanthus* present few problems to the grower. They thrive in short grass or in ordinary well drained garden soil. They will flower best in sheltered places, particularly if the sun can "bake" the corms in summer. They are also ideally suited to being potted up thickly for flowering in a cool greenhouse or indoors. They should be planted in the autumn when first received from the nursery, at least 3 inches (7.5cm) deep, and deeper still in light soils that dry out easily. If they are planted in grass, the grass should not be mown until the leaves have started to die down, in April.

Propagation happens naturally underground, but the process can be assisted by digging up the flowered corms when the foliage is dying down, drying them off for a few days, and planting the small offsets in nursery rows in a sunny place. Eighteen months later they can be dug up and replanted in their flowering positions. Sowing collected seed is possible, but will yield up a variety of different-coloured seedlings, as many *C. chrysanthus* cultivars are hybrids with *C. biflorus*. The process from seed-germination to flowering takes three or four years.

Spring-flowering crocuses are mar-

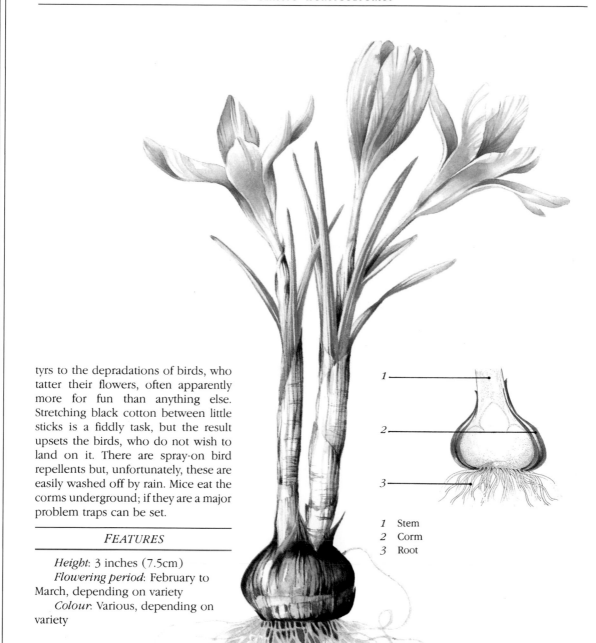

1   Stem
2   Corm
3   Root

tyrs to the depradations of birds, who tatter their flowers, often apparently more for fun than anything else. Stretching black cotton between little sticks is a fiddly task, but the result upsets the birds, who do not wish to land on it. There are spray-on bird repellents but, unfortunately, these are easily washed off by rain. Mice eat the corms underground; if they are a major problem traps can be set.

FEATURES

Height: 3 inches (7.5cm)
Flowering period: February to March, depending on variety
Colour: Various, depending on variety

CYCLAMEN PERSICUM

Sowbread

*C*YCLAMEN PERSICUM is the parent of the large-flowered greenhouse Cyclamen but infinitely more delicate in appearance. It was introduced to Britain from the Eastern Mediterranean (probably Cyprus) in 1731. The Greek word *cyclos* (or *kyklos*) means "circle"; the plant's generic name may refer to the shape of the seedpod, or the spiral the stalk makes as the seedpod matures, or even the rounded tuber. The name Sowbread arose, according to Miller, because "the Root is like a loaf, and the Sows eat it". Certainly wild boars are supposed to have eaten it in southern Europe, but this cannot be said to be one of the more enlightening common names.

The species *C. persicum* is rarely grown in gardens because it is not very hardy but, if a warm corner is found, it will produce its pink or white flowers in March or April. These species are bigger than those of other *Cyclamen* species but with smaller and narrower petals than the modern greenhouse Cyclamen. The leaves are dark green, mainly heart-shaped, and have characteristic "marbling" or "silvering" on them. The flower-stem of this particular species does not spiral as the fruit matures.

CULTIVATION

C. persicum should be planted in a very well drained soil enriched with organic matter, in a warm but not sunny corner where it can be protected by shrubs. Even with these precautions, this species is unlikely to survive indefinitely, except in the most favoured mild districts. The corms cannot be divided so freshly dug-up seeding corms should be planted about an inch (2.5cm) deep in early autumn. These are more likely

to retain some roots, making it easier to identify which is the bottom and which the top of the corm — never an easy matter. Alternatively, pot-grown corms can be bought from some nurseries, which solves that particular problem. *C. persicum* can also be grown in pots containing loam, leafmould, peat and sand in equal parts and kept in a cool greenhouse.

The modern varieties of *C. persicum* are definitely tender and need to be grown in pots in John Innes No. 1 compost in a greenhouse at a minimum winter temperature of 10°C (50°F). When flowering they will need a minimum of 13°C (55°F). If grown as houseplants, their pots should stand in trays of pebbles, which can be watered to counteract the baleful drying effects of modern central heating. When they have finished flowering and the leaves have begun to die down — usually in June — water should be withheld completely. In August, they should be potted up in a slightly larger pot and watered well as they come once more into leaf.

Seed of this species and of modern greenhouse varieties is sown in August or September in pots put in a cold frame. The tender forms are overwintered in the heated greenhouse (or seed can be sown in January under glass), potted up in the spring, and put

into their final pots in summer. With luck they should flower the following autumn. Seedlings of the hardier *Cyclamen persicum* can spend the winter in a protected cold frame and be planted out the following May.

FEATURES

Height: 6-9 inches (15-23cm)
Flowering period: March and April
Colour: Pink
Scent: Yes

1 Petal
2 Ovary
3 Stigma

1 2 3

DAHLIA COCCINEA

*T*HIS was one of the first dahlias to be introduced into Europe from their native Mexico. It went first to the Botanic Gardens at Madrid in 1789, and was named in honour of a Swedish pupil of Linnaeus called Dahl. In 1804, Lady Holland sent seed from Madrid to Mr Buonaiuti, her husband's Italian librarian, who grew it successfully. (It had been introduced to England before, but had not survived.) At the same time it was being cultivated in France, most notably in Empress Josephine's garden at Malmaison; some new cultivars found their way to Britain at the end of the Napoleonic Wars. By 1829 J.C. Loudon was reporting in his *Encyclopaedia of Gardening* that it was "the most fashionable flower in this country, and the extent of its cultivation in some of the nurseries ... is truly astonishing". Joseph Paxton wrote a book about its culture in 1838. The Dahlia became a florist's flower, grown, like the pinks, by the Paisley weavers. Dahlias were at the height of their popularity in the mid-19th century, after which there was a decline, when they were preserved primarily in cottage gardens. This century has seen a revival, which is not surprising considering the Dahlia's late-flowering habit and the usefulness of the flowers for flower arranging. Unfortunately, in the process of development, a great deal of the charm of the original three species has been sacrificed for size and shape of flower-head. Of the many cultivars presently in cultivation, it is thought that only the single-flowered ones have *D. coccinea* blood, the rest being primarily descendants of *D. pinnata* and *D. rosea*. The pure *D. coccinea* is grown only in botanical collections.

The colour range of the flowers of the species *D. coccinea* is from yellow through orange to bright red. The flowers are held on 3-feet (90cm) high stems, which often branch; the leaves are pinnate with narrowly ovate leaflets.

CULTIVATION

Dahlia tubers are planted in the ground in mid-April, at least 4 inches (10cm) deep, for they are extremely frost-tender. If rooted cuttings are taken from shooting crowns in spring, these can be planted about the third week in May. Dahlias of all kinds do best in full sun in a rich, well drained soil, into which bonemeal has been incorporated before planting. They need to be staked sturdily, one thick stake to each plant. They will need watering whenever the weather is dry in summer for they have large, thirsty leaves, and water-stress affects the flowering adversely. About three to four weeks after they have been planted, the stem should be "stopped" to encourage it to branch. Dahlias grown for show should be disbudded regularly, so that there are only a few, but large, flowers.

When the frosts have blackened the foliage in October, the tubers have to be lifted. (They will survive a mild winter out of doors, especially if covered in straw, but who knows for certain when we will get one of those?) The stems should be cut back to about 6 inches (15cm), the tubers carefully dug up, and the whole crown hung upside down for a week to dry out. Once that is done, the tubers can be put in a box, barely covered with moist peat and stored somewhere where the frost will not penetrate. Dahlia tubers shrivel with monotonous regularity in winter and need to be inspected often and when necessary "reinflated" by a night spent in a bucket of water. The easiest

1 Stem
2 Tuber
3 Root

method of propagation is undoubtedly division of the tubers in March, although it is possible to start them into growth in heat in the greenhouse — 15-18°C (59-64°F) — in mid-February or later, and take 3-inch (7.5cm) cuttings from the resulting shoots.

FEATURES

Height: Up to 3 feet (90cm)
Flowering period: August until October
Colour: Yellow, orange, red

DAPHNE COLLINA

*D*APHNES give a great deal of trouble to gardeners, yet we are always ready to cultivate them, even though they often do not thrive at all or suddenly die for no apparent reason. Their neat long-lasting flowers and, above all, their scent are enough to make all but the dullest gardener long to grow them. There is one Daphne, *D. mezereum*, which may be native to Britain or was perhaps introduced so long ago that it appears to be. It is known as 'Mezereon', a corruption of its Latin specific name. It is one of the worst for suddenly dying out without so much as a by your leave.

D. collina is not native to Britain: it comes from the Mediterranean region, from the west coast of Italy (around Naples and north as far as Tuscany) and also from Crete and southern Turkey. Nevertheless, it is hardy in most British gardens, although it is not very long-lived. It has been grown in Britain at least since 1752.

It makes a neat evergreen bush up to about 3 feet (90cm) high in a well favoured place. The leaves are oval and tapered at the base, up to 1½ inches (4cm) long but quite narrow. They are dark green and shiny above, and paler below. The flowers are very fragrant, and are dark pinkish-purple in colour. They are produced in a cluster of about 10 flowers, each one half an inch (1cm) across. These appear in May and continue into June. *D. collina* is often grown in a rock-bed because it is both well behaved and rather slow-growing.

D. collina is one parent of an excellent, although uncommon, offspring called *D. × hybrida*, the result of a cross with *D. odora* made in France in 1820. This flowers practically continuously, except for a brief period in summer, and is very fragrant.

CULTIVATION

Because *D. collina* is not bone-hardy, it is best grown in a well drained spot and in full sun. It should be planted either in September or in March-April. Cuttings of nonflowering sideshoots are usually taken in July and put in a free-draining peat and sand mixture in a cold frame. Once rooted they can be potted up, plunged in the ground for the summer, and planted only when they have reached a reasonable size. If there are low-growing shoots on a bush, these can be layered.

FEATURES

Height: 2-3 feet (60-90cm)
Flowering period: May and June
Colour: Purple-pink
Scent: Very scented

1 Petal
2 Stamen
3 Ovary

DIANTHUS PLUMARIUS

Garden Pink

*D*IANTHUS PLUMARIUS is the Pink, as opposed to the Carnation, and altogether a daintier and more useful garden plant. It is a native of the Alps, of Austria and of Hungary, and is thought to have been introduced to Britain by the Normans with the stone they brought over for castle-building. Certainly this plant can be seen growing in the walls of castles such as the one at Rochester. In the 16th century Pinks were called "Feathered Gillofers" (gillyflowers) or "Soppes in Wine", this last because they were dipped in wine to impart a clove-like scent. Gerard grew them, and is supposed to have named them *plumarius* because of the "feathered" (deeply cut) petals. Parkinson had 17 varieties, and John Tradescant the Elder found them while on a trip to Russia in 1618. The first laced pink saw the light of day in 1772; it was called "Duchess of Lancaster" and was described as "laced" because the edges of its petals were tipped with the same colour as the central blotch. The laced pinks were much grown by the Paisley weavers (indefatigable searchers after new florists' flowers) in the mid-19th century, and also by north-country miners. The Paisley weavers were obsessed by pinks in the period between 1828 and 1850, and in that time they raised more than 300 sorts. In this century, Montagu Allwood has bred a splendid race of pinks called "Allwoodii", which are hybrids between pinks and perpetual carnations, and combine many of the best qualities of both.

On the face of it there seems no connection between the word *Dianthus*, or even "gillyflower", and that of "carnation" — indeed, even when one does know the connection, it seems far-fetched. The Greeks, who called *D. caryophyllus* the "divine flower", made garlands (Latin *coronae*) of it, which is how it is supposed to have got the name "carnation". The specific epithet *caryophyllus* means "nut-leaved", and derives from the similarity between its scent and that of the clove (the "leaved" part refers to the leaves of the tree from which cloves are harvested). It all seems very insubstantial, but there it is.

The descendants of *D. plumarius* are both the "old-fashioned" pinks and the modern pinks, which are forms of *D. × allwoodii*. "Mrs Sinkins" and "Emil Paré" are well known examples of the former, and "Doris" of the latter. All have grey-green leaves and flowers up to 2 inches (5cm) across, and grow up to 15 inches (38cm) high. Their flowers are strongly scented.

CULTIVATION

Garden pinks repay care taken over their cultivation, although they will survive in an ordinary soil without much difficulty. They will languish in a waterlogged soil, however; so, if the drainage is suspect, the soil should be dug and raised and grit should be incorporated to lighten it. They are one of the groups of plants which genuinely grow better in an alkaline soil, so, if the soil is acid, a dressing of lime is advisable. They can be planted in either the spring or the autumn. In the former case, they will require some watering in dry periods but, once well established, are unlikely to need water except in times of drought. A potash feed in spring will aid flower development. Pinks (especially the "old-fashioned" varieties when newly planted) are rather loath to make good flowering side-shoots unless encouraged to do so, so early flowers should be snapped off. After they have flowered, a trim over and potash feed will encourage them to flower again.

Cuttings are easy to root if taken in mid-summer. Take side-shoots, about 3 inches (7.5cm) long, trim them off and put in a cuttings compost in a cold frame. They are easy to layer at the same time; this is done by cutting up from below a node to make a peg of stem and fixing the peg in contact with the soil using a piece of bent wire.

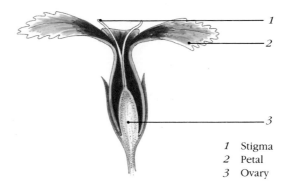

1 Stigma
2 Petal
3 Ovary

FEATURES

Height: Up to 15 inches (38cm)
Flowering period: June and July; also September
Colour: Various; white through to dark red
Scent: Spicy

DIERAMA PULCHERRIMUM

Wand Flower; Angel's Fishing Rods

*T*HIS is a very beautiful, graceful plant, far too rarely grown. It has tufted, grassy leaves, and thin arching stems 6 feet (1.8m) long at most, with rose-purple or red pendulous trumpet-shaped flowers which appear in September and October. It was introduced to Britain by the Backhouse of York nursery in 1865 from the Buffalo River region of eastern South Africa. There is a white form as well as several named cultivars, such as 'Windhover' (a very tall rose-pink variety) and 'Port Wine' (purple), mainly raised by W. Slinger and his son Leslie before the war at the famous, now defunct Slieve Donard nursery, in Northern Ireland. These days it is hard to find any *Dierama*, except the species, mentioned in nursery catalogues.

CULTIVATION

As *D. pulcherrimum* is native to South Africa, one would expect the corms to be unreliably hardy. That is in fact the case, but, except in very cold localities, it is safe to leave them in the ground through the winter provided some sort of blanket protection can be given. In very cold places the corms should be lifted in November and stored as one would *Dahlia* or *Acidanthera*, then replanted deeply in April. Dieramas like a well drained but fertile soil, enriched with leafmould or compost, and are happiest in full sun — although they will tolerate, and flower in, some shade. They can be increased by sowing seed in spring and planting out the seedlings in the autumn; or the corms can be lifted and divided in October and the small offsets planted in a nursery row to grow on for a year before being planted where they are to flower. If possible, it is best to leave them in peace if they are to flower every year.

FEATURES

Height: Up to 6 feet (1.8m)
Flowering period: September and October
Colour: Purple or deep red

1 Ovary
2 Stamen
3 Petal
4 Style

DIGITALIS PURPUREA

Foxglove; Witch's Thimble; Finger-flower

DIGITALIS PURPUREA is native to Britain and has been used in the treatment of epilepsy for hundreds of years. The discovery of digitalin's usefulness for heart complaints, however, is of comparatively recent origin. Dr William Withering first used it in the control of dropsy at the end of the 18th century, but it was not until a little later that its efficacy as a heart stimulant was appreciated. It is one of the few native British plants used in modern medicine. The suffix "glove" is thought to stem from the Saxon word for "bell", although it seems as likely to come from the word for a hand-covering, considering that *Digitalis* means "finger-flower".

In good conditions it is a perennial plant, although most people consider it a biennial because it will not flower the first year after germination, instead overwintering as a rosette of large, oblong leaves in true biennial style. It is a plant of the cottage garden, free-seeding and sometimes a nuisance, but its purple or white flowers (spotted inside the "glove") associate well with Old Roses, and it will flower (from the bottom of the stem upwards) all summer. In a good soil it will reach 5 feet (1.5m) in height. If it is desired that the plants be perennial, their seed-heads should be swiftly removed to prevent them wasting energy in seeding.

The species is the least attractive of the foxgloves, easily eclipsed by the 'Excelsior' strain, whose flowers are held so that the markings may be seen easily and whose colours range from white to cream, pink and purple. *D. ambigua* is a perennial with yellow-brown flowers, which are intriguing, but the best is *D.×mertonensis*, a perennial with long-lasting strawberry-coloured tubular bells.

CULTIVATION

Cultivation of this plant is extremely easy; indeed, most people find that, within the confines of a tidy, orderly garden, its prodigal seeding makes its presence a decidedly mixed blessing. It is happiest in semi-shade and in ground which does not become too dry; it can survive in dry conditions but will not grow to its optimum height. It is very easily propagated by seed sown thinly outside in May, with the seedlings being transplanted first to a nursery row and then to their flowering positions in September. This plant is more than a match for any common or garden pest or disease.

FEATURES

Height: Up to 5 feet (1.5m)
Flowering period: June until September
Colour: Purple or white, with dark spotting inside

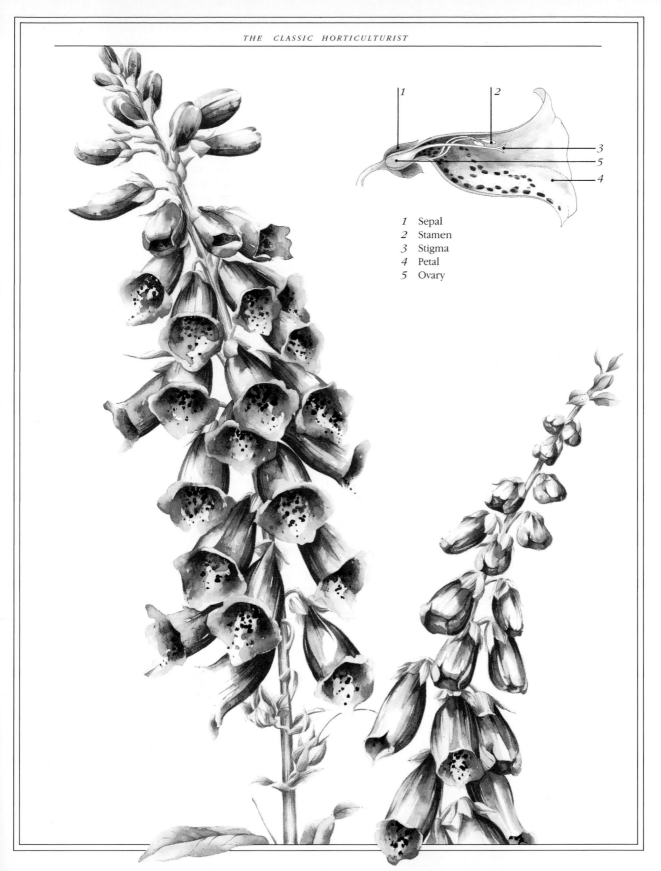

1 Sepal
2 Stamen
3 Stigma
4 Petal
5 Ovary

FRITILLARIA IMPERIALIS

Crown Imperial

*F*RITILLARIA IMPERIALIS, the Crown Imperial, is a native of Iran and of the Himalayas. It is one of those plants which, as a result of being cultivated for a very long time, has acquired accretions of legend and myth. There is an ancient Persian story that a wife wrongly accused of unfaithfulness was turned into this flower, and that she will continue to weep until such a time as she is united with her husband once more. Her "tears" are the nectaries which can be seen at the base of the flowers. E.A. Bowles recounts another story, that the Crown Imperial (which was then white) was growing in the Garden of Gethsemane and was too proud of its looks to bow its head when Jesus came. Since then, it has hung its head and blushed for shame with tears in its "eyes".

This plant was cultivated in Turkey for a long time before it was introduced to Vienna in 1576 by the botanist Clusius. It had certainly arrived in Britain by 1590. Gerard wrote of it:

This plant hath been brought from Constantinople, amongst other bulbous roots, and made denizons in our London gardens, whereof I have great plenty.

It is a very noble plant, which has only one drawback — namely, that all parts of it smell strongly of fox. Apparently it used to be eaten in Persia, but only after being boiled until the evil smell (and poisonous properties) disappeared. Nevertheless, the Persians must have had strong stomachs! Bowles likened the smell to

a mixture of mangy fox, dirty dog-kennel, the small cats' house at the Zoo, and Exeter Railway Station, where for some unknown

reason the trains let out their superfluous gas to poison the travellers. [My Garden in Spring, 1914.]

The plant will reach about 3 feet (90cm) tall in good conditions. It has lanceolate fleshy green leaves which are carried in what are called "whorls" to halfway up the thick, dark stems. The flowers, below a topknot of leaves, hang, are about 2 inches (5cm) long, and bloom in April. They are usually orange in colour, although there are yellow or red varieties such as *F. maxima lutea* and *F. maxima rubra*. After the flower is fertilized, the seeds begin to form in a large fruit which gradually turns upwards.

CULTIVATION

Once planted, these bulbs are happiest left alone to flower, which they do for many years without degeneration. They prefer a position in full sun, but will tolerate a certain degree of dappled shade. The bulbs are very easily damaged by careless handling, and you are wise to wear gloves when planting them, as much to protect yourself from the smell as to protect the plant from harm. They should be planted as deep as can be managed — at least 8 inches (20cm) deep if they are to flower consistently well and, because they have hollow centres, they should be planted on their sides to prevent them from filling with water. They can be potted up and flowered in a conservatory, but as they are perfectly hardy this is not strictly necessary. Seed of *Fritillaria imperialis* can take up to six years to flower; more satisfactory is to remove the offsets, pot them up and put them in a frame or plant them in nursery rows.

FEATURES

Height: 3-4 feet (90-120cm)
Flowering period: April and May
Colour: Pale orange
Scent: Strong and unpleasant

1 Ovary
2 Nectar drops
3 Stamen
4 Stigma
5 Petal

GALANTHUS NIVALIS

Common Snowdrop; Milk Flower

*T*HE well known Common Snowdrop, *Galanthus nivalis*, is probably not a native of Britain, although both the single and double forms have become naturalized in some places. It is definitely indigenous to France and to Eastern Europe as far as the Caucasus. Some people believe it was introduced to Britain from Italy by monks in the 15th century for use during the Feast of the Purification of the Virgin (or, as we better know it, Candlemas). During this festival, the image of the Virgin Mary would be removed from its usual resting-place and snowdrops put there in its stead. The snowdrop, therefore, became a symbol of purification. Gerard mentions it, calling it the "Early Flowering Bulbous Violet" (everything was a violet in those days — unless it was a lily!). The plant has a documented history much older than that, however, for in 300BC Theophrastus mentioned that the snowdrop was growing on Mount Hymettus.

The snowdrop has a special place in our hearts because, with the exception of the Winter Aconite, *Eranthis hiemalis*, it is the first small flower of the year to appear. It never misses a season, being utterly dependable and completely hardy. The cruellest winter weather cannot sully the flowers. The flower stem grows to 7 or 8 inches (18-20cm) tall, although it can be much shorter, especially if the bulbs are crowded. The flower is about an inch (2.5cm) long; the inner segments are streaked with green. The flowers are faintly scented of honey. There are usually two leaves, which have a pronounced keel; they are linear and bluey-green in colour. There are many varieties of this plant, one of the most unusual being *G.n.* var. *scharlokii*. This variety has a pair of leafy spathes which stand up above the drooping flower head. As Bowles put it in *My Garden in Spring* (1914): "In *G. scharlokii*, these queer little leaves stand up and spread out over the flower with an expression like that of hares' ears." Other forms include a yellowish snowdrop called *G. lutescens* and one with the outer segments tipped with green, called *G. viridapicis*. The most common variant is the double form, called *G. flore pleno*, but it is fat and gawky and hardly an improvement on the single flower. In some years *G. nivalis* will flower in January, but it is more usual to see it in February and early March.

CULTIVATION

Snowdrops have the most agreeable trait of thriving best if left alone or if transplanted "in the green". This means that they are best moved just after flowering, when one can see them and when one thinks about it. Division is needed only when the clumps are *really* overcrowded. Once you know this much about snowdrops there is little more you need to know. They grow anywhere, provided that the soil does not dry out, which means they are happy on heavy loams and do not seek the hot dry places which so many good plants seem to require. They like positions under deciduous trees and will grow in short grass, and their leaves are dying before the first grass cut is due, so they are never a nuisance. Seed can be sown anytime between October and March in a cold frame but it will be several years before the seedlings flower. *G. nivalis* is little troubled by pest or disease.

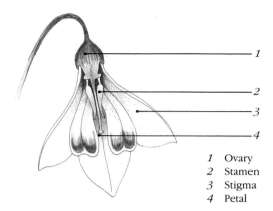

FEATURES

Height: 3-8 inches (7.5-20cm)
Flowering period: February and March (January in mild seasons)
Colour: White with green streakings on inner petals
Scent: Faintly of honey

1 Ovary
2 Stamen
3 Stigma
4 Petal

GARRYA ELLIPTICA

The Quinine Bush; The Silk Tassel Bush; Fever Bush

*G*ARRYA ELLIPTICA is in its own family, Garryaceae. It was found in 1828 by the famous explorer and plant hunter David Douglas near the sea on the south side of the Columbia River in Oregon; it is also a native of California. He sent it home to the Horticultural Society; it flowered first in England in the Society's garden at Chiswick in 1834. It had actually been found some years before by Archibald Menzies, while on an expedition with Captain George Vancouver between 1790 and 1795, but it was not named by him. Douglas named it after the then deputy governor of the Hudson Bay Company, Nicholas Garry, as a mark of gratitude — the employees of the Hudson Bay Company were very helpful to him on his travels, providing him with guides and supplies. (Poor Douglas came to a sticky end in Hawaii in 1834, at the age of only 35, when he fell into a trap which already contained a wild bull. His introductions include many of our best-loved conifers, such as the Douglas fir, *Pseudotsuga taxifolia*, as well as the flowering currant, *Ribes sanguineum*, and the Oregon grape, *Mahonia aquifolium*.) *G. elliptica*'s common names of "The Quinine Bush" and "Fever Bush" arise because a decoction of the bark and leaves was used in North America to treat fevers.

With *G. elliptica* the two sexes are on different plants; the male, which is the showiest, was the first to be introduced to Britain, and the female followed some years later. The male has much the longer tassels of flowers and is the more garden-worthy. *G. elliptica* is an evergreen shrub, or occasionally small tree, with oval leaves which are green above and grey below. The flowers are in hanging catkins, some-times up to a foot (30cm) long on the male in sheltered, warm situations. They are grey-green in colour. The female flowers are usually about 3 inches (7.5cm) long and more silvery in tone. This plant flowers in February and March — even earlier in mild seasons. The fruits, which are produced only when there are male and female plants growing near to each other, are round, silky and a greenish-purple in colour.

CULTIVATION

The most important thing to remember about the cultivation of garryas is that they do not like to be moved; most especially, they do not like their roots disturbed. They are, therefore, usually to be found in nurseries in containers and consequently can be planted at any time of the year, provided they are adequately watered. However, late spring is the best time to plant them. As they are not bone-hardy, they should be protected for the first winter or so until the young wood has had an opportunity to toughen up. Garryas have the happy knack of surviving very well on a north wall, but they will flower more generously on a west or south one. Semi-hard cuttings can be taken with a heel in late summer and rooted in a cold frame. Alternatively they can be layered in autumn. Nothing much harms them, although severe frost will brown the tough leaves.

FEATURES

Height: Up to 12 feet (3.7m)
Flowering period: November to February
Colour: Silver-green

1 Petal
2 Stamen

HELLEBORUS NIGER

Christmas Rose

OVER the years, hellebores have attracted to themselves a wealth of superstition and story. This is hardly surprising when one considers that they are poisonous (like so many of the Ranunculaceae) and yet, in small quantities, medicinally efficacious — for example, the alkaloids they contain have been used in the past as a drastic laxative and as a heart stimulant. The Greeks thought that hellebores could be used as a cure for madness, and this belief lingered on into the 17th century. Gerard, for example, reported that a 'purgation' of hellebore 'is good for mad and furious men, for melancholy, dull, and heavie persons, and briefly, for all those that are troubled with black choler and molested with melancholy'. To this end, the roots were dried, ground down, and taken like snuff. They were also thought to be useful for warding off evil spirits. The Christmas Rose even plays a part in a medieval mystery play. A country girl accompanying the shepherds cries because she has nothing, not even flowers, to bring to the Baby Jesus. An angel takes pity on her, touches the ground, and up flowers the Christmas Rose.

H. niger, being a *bona fide* cottage-garden plant, has enjoyed a good press from, for example, Gertrude Jekyll and Vita Sackville-West. The latter wrote:

> *I have a plant in my garden which to my certain knowledge has been there for fifty years. It was bequeathed to me by an old country-woman of the old type, who wanted me to have the enjoyment of it after she had gone.*

H. niger is so-called because its roots are very black. It comes from Central and Southern Europe, as well as western Asia, and arrived in this country in the 16th century. It is an evergreen perennial with dark green leathery leaves deeply divided into 7 or 9 segments. The veins on the leaves are very pronounced. The pure white saucer-shaped flowers, up to 2 inches (5cm) across and with yellow stamens, come out in succession between December and March. There are some good forms which have been raised, or geographic variants which have been introduced, one of the best being the larger-flowered 'Potter's Wheel'. The Lenten Rose is *H. orientalis*, which appears from February until April, with hanging flowers of white, cream, pink or purple. *H. viridis* and *H. foetidus* are rare British natives.

CULTIVATION

The Christmas Rose is one of the those plants about which gardeners are inclined to be wary, and rightly so. The reason is that it will thrive in some gardens but not in others, and there seems little rational explanation for why this should be so. The best hope of success lies in giving it a place in partial shade and in a soil which is rich and moist, but not waterlogged. Plants should be left sternly alone and not transplanted; the only attention they appreciate is a top-dressing of well rotted compost or manure in October or after flowering; this acts also as a much needed mulch

in summer. The precaution is often taken of placing a cloche over the open flowers to stop them being spoiled by mud-splashes, as they flower so early in the year; this looks unsightly but does the trick. The flowers can be cut in bud and brought inside where, if their stems are split,they will last quite well in water. If the plant must be divided, this should be done soon after flowering in March. Division is the usual way of increasing the plant, although it is possible to sow ripe seed in early summer in a seed tray which is placed in a cold frame (hellebores are very hardy). Once germinated, the seedlings can be planted in a nursery bed until large enough to plant out.

FEATURES

Height: 12 inches (30cm)
Flowering period: January and February (December in some years)
Colour: White

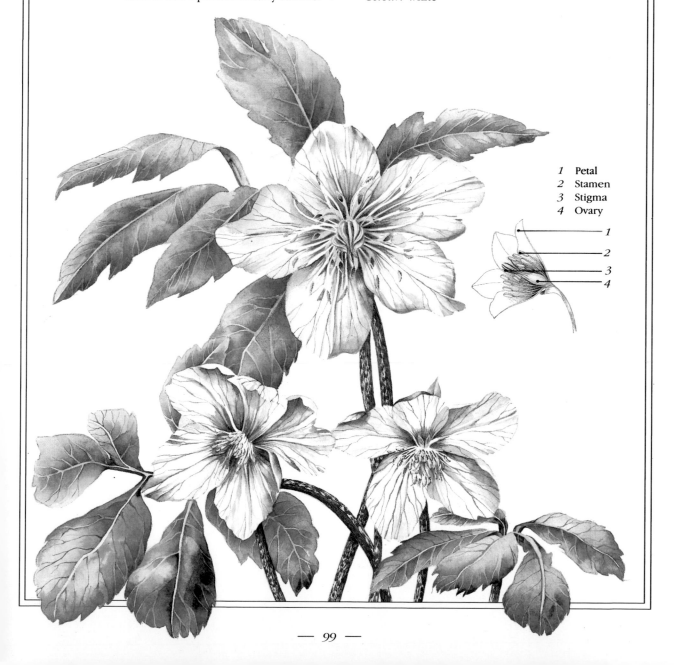

1   Petal
2   Stamen
3   Stigma
4   Ovary

IRIS GERMANICA

Purple Flag; London Flag

THE goddess Juno had a messenger, called Iris, whom she regularly sent down to Earth. Iris reached Earth by way of a rainbow bridge, and the Iris flower is named after her because it is found in such a rainbow of colours. *Iris germanica* was for long thought to be one of the parents of the modern garden Bearded Iris hybrids but, in fact, these are primarily a mixture between *I. pallida* and *I. variegata*. However, *I. germanica* is the oldest cultivated European iris, having been grown at Lake Constance in the 9th century. It grows in the wild all over Europe and eastwards to India and Nepal. It is likely that it has been grown in gardens in the Middle East and the Mediterranean for centuries for its medicinal and cosmetic properties. In Italy, the rhizome was even used to make rosary beads. A subspecies is the famous *I. florentina*; this iris is the source of orris root, so popular in the Middle Ages and later, and was grown extensively around Florence for use in the preparation of beauty treatments and toothpastes. The roots of this iris, when fully dry, are very scented, which makes it ideal for perfuming toilet waters and for putting among household linen. The scent is similar to that given off by violets.

I. germanica and *I. florentina* are not the French Fleur-de-Lis. The honour for that must go to *I. pseudacorus*, the Yellow Flag Iris. The story goes that in the 6th century Clovis I, one of the Merovingian kings, was being pursued by the Goths near Cologne. He saw an iris growing in the middle of the Rhine and deduced rightly that the water must be sufficiently shallow for his men to cross in safety. He consequently adopted *I. pseudacorus* as his emblem. Six centuries later, Louis VII of France copied him and wore the emblem while fighting in the Crusades, with the result that this iris acquired the common name Fleur-de-Louis, soon shortened to Fleur-de-Lis.

I. germanica grows up to 3 feet (90cm) tall. It has greenish-blue leaves and scented flowers, the falls of which are purple, the beard yellow, and the standards mauve. It flowers in May. *I. florentina* is a form of it; the flowers are white with a touch of blue on the falls.

CULTIVATION

Rhizomatous irises, like garden pinks, are happiest on a neutral or slightly alkaline soil. They like a soil into which well rotted organic material and bonemeal have been incorporated. They are best planted soon after they have flowered in early July, or in September. The rhizomes like to be "baked" by the sun in summer, so should be planted shallowly with the longest side facing the sun; they should not be shaded by other plants or weeds. In planting, their long "fans" of leaves should be cut back by half with a sharp knife so that the plant does not lose too much water by transpiration before the roots are established. *I. germanica* needs to be dug up and divided every three years, to prevent the rhizomes becoming overcrowded and ceasing to flower freely. Division means throwing away the old central part of the rhizome and replanting only the younger branches; these can be divided by snapping them apart or using a sharp knife. The new transplants should be kept well watered until established, after which time watering will probably be unnecessary. A general balanced fertilizer sprinkled around the plants in March helps their development.

FEATURES

Height: 2-3 feet (60-90cm)
Flowering period: May
Colour: Bright purple with yellow "beard"
Scent: Yes

1 Style
2 Stigma
3 Stamen
4 Ovary

KNIPHOFIA GALPINII

Galpin's Red Hot Poker; Torch Lily

*T*HIS is a dwarf Red Hot Poker, growing not more than 2 feet (60cm) tall, with stiff, narrow, grassy leaves and loose, flowered spikes which are orange but which fade a little as they age. It was introduced into Britain in 1930 and was immediately recognized as a good garden plant, because of its compact habit; it has since been the parent of several good garden cultivars. It was named after one Ernest Galpin, of Barberton, who had sought plants in South Africa in the 1890s. This species comes from the Transvaal, Swaziland and Natal. It is plainly very closely related to *K. macowanii* and *K. triangularis*; the *K. galpinii* of gardens may well be a subspecies of the latter. The work on hybridizing has been done by Amos Perry, Alan Bloom at Bressingham, the Slingers at Slieve Donard, and Beth Chatto. Among the best cultivars for gardens are 'Little Maid' and 'Sunningdale Yellow'

Kniphofias, or *Tritoma* as they were called in the 19th century because many species have leaves with three edges to them, have always been popular border plants, even though the usual one seen, *K. uvaria*, is rather coarse and unsubtle in colour. In *The Wild Garden* (1870) William Robinson recommended that tritomas be planted in bold clumps, and Gertrude Jekyll grew them for their strong colour and imposing shape in September in her main flower border at Munstead Wood.

CULTIVATION

Kniphofias are amenable plants which need little or no feeding and desire only, like Greta Garbo, to be left alone. Being South African, however, they do not appreciate wet feet, and so are best planted in full sun and in a light, free-draining soil. In exposed or cold places some winter protection — straw, perhaps — is a help, especially when the plant is young. They should be planted either in September or in April, the most suitable times for planting not-completely-hardy plants. The roots are fleshy and should not be allowed to dry out before planting. Red Hot Pokers do not need staking; they are quite capable of holding their heads up on their own. Propagation is done in April, by division or by seed sown outside in rows.

FEATURES

Height: 2 feet (60cm)
Flowering period: September, October
Colour: Flame and orange-yellow

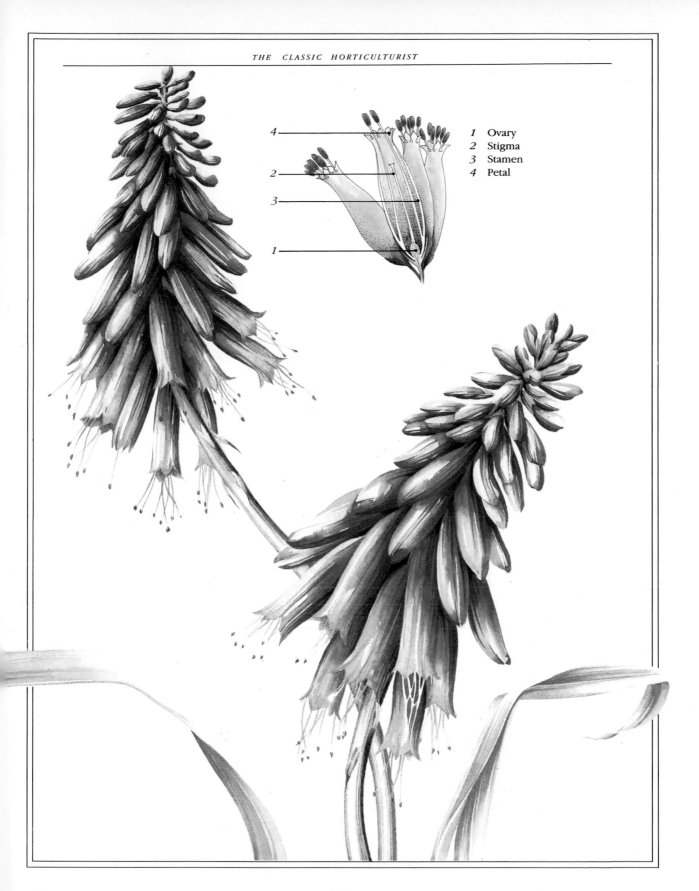

4
2
3
1

1 Ovary
2 Stigma
3 Stamen
4 Petal

LILIUM CANDIDUM

Madonna Lily

LILIUM CANDIDUM is the Madonna Lily. It is thought to be the oldest cultivated plant, although that is a theory very hard to prove — or disprove — conclusively. The Phoenicians knew of it, as did the Assyrians, and the Minoans of Crete painted it on their vases before 1600BC. It is likely that it was as much prized for its manifold medicinal qualities, which included cures for dropsy and for boils, as it was for its considerable beauty. The Roman army, for example, planted it near their permanent camps because of its supposed efficacy in curing corns, a condition from which such a mobile army must have suffered badly. It was probably the Romans who introduced it into Britain, perhaps from Salonika. It finds a place in Ion Gardener's early-15th-century *The Feate of Gardening.*

It is a hardy lily which grows up to 5 feet (1.5m) in height. The flowers, several to a stem and the purest white except for the yellow pollen, are trumpet-shaped, with the petals reflexed at the ends. They are 3½ inches (9cm) long and very, very fragrant. They flower in June and July. There is a variety called *L.c. cernuum,* which has narrower petals, and a form of this is called 'Sultan Sambach'.

CULTIVATION

L. candidum is not the easiest lily to grow, mainly because it is a martyr to *Botrytis* as well as lily virus. The first can be minimized by planting the lilies in a sunny place and spraying with Bordeaux mixture in warm, wet seasons. The second, which manifests itself as mottling and streaking on the leaves, requires that the infected bulb be dug up and burned to prevent its spread to other bulbs.

The bulbs should be planted shallowly, preferably in a slightly alkaline soil, with the top of the bulb just below soil level, no later than the beginning of September, and thereafter left undisturbed. The soil should be enriched and, if it is heavy, lightened by digging in grit. As *L. candidum* is a species, its seeds will breed true and can be sown deeply in pots in early autumn and put in a cold frame. Alternatively, the bulbs can be uncovered in the autumn, the outer scales removed, half-buried in trays of cuttings compost, and put in a cold frame to produce little bulbs. When this has happened, they can be potted up in potting compost and plunged in the open ground to grow to flowering size.

FEATURES

Height: 4-5 feet (1.2-1.5m)
Flowering period: June and July
Colour: Pure white
Scent: Very strong and sweet

1 Stigma
2 Stamen
3 Ovary
4 Petal

MAGNOLIA × SOULANGIANA

*M*AGNOLIAS are the aristocrats of the garden; their lineage is ancient and their appearance most distinguished. Their line stretches back into the distant mists of time. It has been proved that they grew 5 million years ago, which makes them as old as even the Maidenhair tree.

Magnolia × soulangiana is a hybrid between *M. denudata* and *M. liliiflora*. This cross took place by chance in the garden of the château belonging to M. Soulange-Bodin at Fromont outside Paris. M. Soulange-Bodin was the founder of the National Horticultural Society of France, so plainly he was a dedicated and knowledgeable gardener. Of this magnolia's parents, *M. liliiflora* is a Chinese plant which was in fact introduced to Europe from Japan in 1790, and *M. denudata* is another Chinese magnolia which came in the year before and which is more usually known by its common name of Yulan. The result of the cross first flowered in 1826 and was soon distributed in England. John Loudon wrote in 1829 that a nursery (Messrs Young of Epsom) "have bought the entire stock of *M. × soulangiana* from M. Soulange-Bodin for 500 guineas in consequence of which that fine tree will soon be spread all over the country". For the stock to sell for such a high price there must have been many plants, even taking into account the novelty value. A French nurseryman called Cels raised some other seedlings from this cross, the most notable being 'Norbertii' and 'Alexandrina'. There are now many forms of *M. × soulangiana* ranging from completely white to claret-purple in flower colour.

The honour of having his name perpetuated in this glorious genus went to Pierre Magnol. He was a botanist in Montpellier in France who never gained promotion while he was a Protestant. However, in 1694, he converted to Catholicism and this opened the way to his preferment, not only as a professor but also as Director of the Montpellier Botanic Gardens. In those days, religious divergence from state orthodoxy was a recipe for professional suicide.

M. × soulangiana makes a large spreading shrub. It is deciduous and has green leaves, up to 6 inches (15cm) long and tapering towards the apex; they are narrowly ovate or lanceolate, shiny above and downy below. The flowers, which appear in April before the tardy leaves, are large and upright at first, later opening out as they mature. They are white but have pink staining at the base. The shrub continues to flower, although less impressively, even after the leaves have developed. One of the best varieties is called 'Lennei'. This originated in Italy but in 1854 was brought to Germany by a nurseryman who called it after Herr Lenné, the man in charge of the Royal Gardens in Berlin. It has rosy-purple flowers.

CULTIVATION

M. × soulangiana is quite the easiest of a difficult but very beautiful genus to grow, which is why it is so widely planted. For one thing, there is little doubt of its hardiness (it does not require the comfort of a warm wall) and, for another, it will grow in any ordinary soil which is not too shallow. As with all magnolias, however, late frosts will play havoc with the flowers, so it is a good idea not to plant it in a hollow or in a place exposed to cold winds or early morning sun. If its preferences can be met it will grow reasonably swiftly, flower quickly (for a magnolia), and not be an anxiety or trouble as far as pests or diseases are concerned. It does respond positively to a mulch of well rotted compost in April. The roots are fleshy and so must be carefully planted; this is an operation best attempted in March or April.

Propagation by seed takes a long time and is anyway uncertain as this magnolia is a hybrid. Layering in spring is preferable. Cuttings can be taken in July if a propagator with bottom heat is available, but nothing will induce this magnolia to "strike" otherwise. Magnolias should not be pruned if it can be avoided.

FEATURES

Height: Up to 15 feet (4.6m)
Flowering period: April to June
Colour: White with pink or purple staining
Scent: Yes

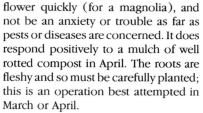

1 Stigma
2 Stamen
3 Petal
4 Ovary

MECONOPSIS GRANDIS

*G*ARDENERS have always cared passionately about the Himalayan blue poppies for a combination of reasons: the difficulty of rearing them successfully, the astonishing beauty of the good forms, and the stories attached to their introduction.

Meconopsis grandis comes from the high meadows of western Sikkim and from Nepal and Tibet. It was first discovered in Sikkim by Sir David Prain in 1895, and the Nepalese form was sent back from Bhutan but a pair of plant collectors called Frank Ludlow and George Sherriff who made seven expeditions in the Himalayas between 1933 and 1947. (In all that time they never called each other by their Christian names!) *M. grandis* has green lanceolate leaves, with reddish bristles. The flowers, up to 6 inches (15cm) across but usually much less, are saucer-shaped and either purple or deep sky-blue, with yellow stamens which make a striking contrast. The flowers come out in May and June.

It is by no means the only garden-worthy meconopsis. Gertrude Jekyll grew *M. napaulensis* (syn. *M. wallichii*) in her woodland garden at Munstead Wood. This has red, purple or white flowers in June and grows up to 6 feet (1.8m) tall. This species is a most imposing plant but, because it is monocarpic, needs to be repropagated after it has flowered. *M. betonicifolia* is, like *Primula florindae* (see page 132), one of Frank Kingdon-Ward's marvellous introductions. Some 40 years after Abbé

Delavay had sent a herbarium specimen back to Paris, Kingdon-Ward found it in the Tsangpo gorge and despatched seed home. It has proved an easy plant for amateur gardeners — which is more than can be said for the other Himalayan poppies.

CULTIVATION

Meconopsis, with the exception of that pretty weed, the Welsh Poppy, demand the impossible. To be truly content they require a great deal of water in the summer as they flower, and hardly any in the winter; the British climate works roughly the other way round. So they do need siting in a semi-shaded place in a soil which does not dry out in summer easily and where, if necessary, cloches can be used to keep them dry in winter. *M. grandis* is perennial and can be increased by detaching and potting up side-shoots when these are made. It also comes true from seed, which can be sown in August or September in a cold frame. Seedlings have to be carefully watched as they can fall victim to downy mildew. *M. betonicifolia* becomes a short-lived perennial, rather than an annual, if it is prevented from flowering the first year by cutting off the stems before they have the opportunity to produce flowers.

FEATURES

Height: 2 feet (60cm)
Flowering period: May and June
Colour: Deep blue or deep mauve

1   Petal
2   Stamen
3   Stigma

MYOSOTIS ALPESTRIS

*A*LTHOUGH it seems cruel to dispel the layman's romantic notions, the experts would cast doubt on the identification of *Myosotis* as the original Forget-Me-Not. Some say the honour should properly go to the Germander speedwell. It may be that, in the past, one plant was meant by the name Forget-Me-Not in Britain and another by *Vergiss-Mein-Nicht* in Germany. The most that can be said is that *Myosotis* seems now to have stabilized as *the* Forget-Me-Not and, that being the case, we might as well assume that it was always so. Henry Bolingbroke, later Henry IV, took the Forget-Me-Not (whatever it was) as his emblem, and the "S" (for *Soveignez* or *Souveraine*) came to be worn by those with Lancastrian allegiances as a sign or badge of their affiliation.

M. *alpestris*, although a native of Western Europe, was not grown as a garden flower until the 19th century. It was not until the 1870s that it became an important constituent of spring bedding schemes, associating particularly well with garden tulips. It has been the parent of many hybrids used for this purpose, notably 'Carmine King', 'Royal Blue', 'Blue Ball', and 'Alba'. The type plant has hairy stems and oblong-lanceolate leaves (which gives point to the name *Myosotis*, or "mouse-ears"), and delicious azure blue flowers with contrasting yellow eyes. It does not grow above 8 inches (20cm) tall, which is why it has always proved so useful as an edger to borders, but which rather undermines its effect when combined with tall tulips.

CULTIVATION

M. alpestris could not be easier to cultivate. Indeed, there are some who say it is too easy and takes liberties, spilling its seed with cheerful abandon all over tidy flower beds. It is a perennial, although it is usually treated as a biennial, sown in June in a seed bed, planted out in September as part of a spring bedding scheme, and dug up to make way for summer bedding the following May. Forget-Me-Nots do not like the conditions to be too dry or densely shaded, but otherwise they are very tolerant.

FEATURES

Height: 3-8 inches (7.5-20cm)
Flowering period: April to June
Colour: Azure blue with yellow eye
Scent: Yes

1 Stigma
2 Stamen
3 Petal

MYRTUS COMMUNIS

*F*OR such a well known plant, it is surprising that *Myrtus communis* is not widely grown. It must owe its unpopularity in Britain to its not being entirely hardy and so having to be grown against a wall, except in the south and west. There is a great deal of folklore and tradition attached to the Myrtle. It was very important, for instance, in Roman mythology and the Romans may well have introduced it into Britain when they colonized the country in the 1st century AD. It is a plant which originally comes from the western part of Asia — Iran and Afghanistan — and it is certain that it was growing in Britain by the 16th century. The Roman goddess of love, Venus, wore a crown of it when she rose from the sea, and this connection with love has led to the tradition whereby Myrtle is included in a bride's wedding bouquet. By the 1770s there were at least 19 varieties grown in Britain and it was important as a cut flower. It was often grown outside in summertime in a tub and brought back under glass for the winter.

Myrtle is an evergreen shrub with very fragrant white flowers and leaves which are scented if crushed. The berries are black and round, and were certainly used by the Romans in cooking. The leaves are ovate and pointed, up to 2 inches (5cm) long, very dark green and shiny above but paler underneath. The flowers grow singly, on stalks; they arise from the axils in the leaves and have many prominent stamens. They come out in July and August, a time when good flowering shrubs (besides roses, that is) are not plentiful. There is a double form and also a variegated-leaved myrtle.

CULTIVATION

Myrtles cannot be said to be hardy in Britain. They will limp out an existence in the south but, to be happy and to grow well, even there they will require wall protection and elaborate blanketing as shelter from the stormy blasts. They are, however, perfect subjects for the conservatory, where they can be grown in a border, or in a large pot or tub which can be brought outside for flowering in summer. If they are to be grown in a 10-inch (25cm) pot they will need John Innes No. 3 compost and regular feeding with a liquid fertilizer in the growing season. Myrtles are increased by short cuttings of lateral shoots taken in summer; these will root in a heated propagating case. Apart from winter protection, myrtles need little or no attention except the removal of their frosted shoots, if any, in spring. Their tough leather leaves are a match for aphids and other pests.

FEATURES

Height: 15 feet (4.6m)
Flowering period: July
Colour: White
Scent: Both flowers and foliage are aromatic

1 Stamen
2 Stigma
3 Petal

NARCISSUS PSEUDONARCISSUS

Lent Lily; Wild Daffodil

*T*HIS is Britain's only native Daffodil. It was originally thought that only the short-trumpeted flowers of *Narcissus poeticus* were real daffodils, or 'affodyls' (a name also given to the white asphodel) — hence the specific name *pseudonarcissus*. Since the Lent Lily is native this does seem rather unfair! Turner writes about it, as does Gerard, who reports it growing in London gardens 'in great abundance':

> *The flower groweth at the top, of a yellowish-white colour with a yellow crown or circle in the middle and flowereth in the month of April or sometimes sooner.*

N. pseudonarcissus has long, glaucous leaves and a drooping, trumpet-like corona of inner petals as long as, but darker in colour than, the spreading perianth petals. It flowers in March and April and can be found wild, although its distribution is rather local, being restricted to damp woods and grassland.

The Lent Lily has had something to do with the breeding of all the long-trumpeted garden daffodils. Among its most famous offspring are 'Golden Harvest', 'Queen of Bicolors', and 'Mount Hood'.

CULTIVATION

This plant, and its garden hybrid descendants, are most useful for naturalizing in grass or among shrubs in borders. Indeed, they will grow anywhere with no fuss, provided they are left alone until the leaves start to yellow and die down. They do best when planted in organically enriched soil which does not dry out, because they may not flower the next year if they do not get enough moisture in the spring after flowering. They are, after all, plants of damp meadows. They should be planted early in the autumn, immediately after they are received, so that they have time to make their roots before the winter. In grass this will require the use of a special "bulb planter" or a spade to lift the turf. They should be planted very deeply in cultivated borders in case they are damaged by forking, and at three times the depth of the bulb in grass.

Most people leave the bulbs alone for many years, especially those grown in grass. Those grown in borders will become congested earlier and will benefit from being lifted and divided every few years. This should be done in July. Sowing by seed should be avoided: it is painfully slow and the results are unimpressive.

Narcissi are victims of various virus diseases. If mottling, streaking, or serrating of the leaf-edges is evident, the affected bulbs should be dug up and burned. This seemingly harsh measure is the only way of containing the virus.

FEATURES

Height: 8-14 inches (20-36cm)
Flowering period: March and April
Colour: Pale yellow
Scent: Yes

1 Petal
2 Stigma
3 Stamen
4 Ovary
5 Spathe

NICOTIANA ALATA (N. AFFINIS)

Tobacco Plant

*T*HIS is the ornamental Tobacco Plant, and should not be confused with the tobacco plant of commerce, which is *Nicotiana tabacum*. The genus was named after the French Ambassador to Portugal, Jean Nicot, who planted tobacco in the embassy garden in Lisbon in 1560 and introduced it to France. Sir Walter Raleigh brought it to Britain in 1585. The ornamental tobacco plant, *N. alata* var. *grandiflora*, comes from southern Brazil, where the leaves are smoked and chewed by the natives. As it has to be treated in Britain as a half-hardy annual, it became popular as a constituent of bedding schemes in the 19th century, when it was often associated with heliotropes. Its advantage, from the decorative point of view, over other types of tobacco plant is that it has less expanse of leaf. It usually grows to about 2 feet (60cm) high, and all parts are sticky. The lower leaves are ovate and about 6 inches (15cm) long; the higher ones on the stem are a little smaller. The flowers of the species, which are borne on loose racemes, are 3 inches (7.5cm) long, greenish-yellow, and very fragrant at night.

Modern varieties have been bred with flowers coloured from crimson through pink to white and green which will stay open during the daytime. Examples include 'Domino' and 'Nicki'.

CULTIVATION

N. alata is not difficult to grow, as long as it is appreciated that it is a half-hardy annual and requires to be sown in heat in March, pricked out into trays, hardened off, and planted out when there is no longer any danger of frost. The planting position should be in full sun and in a fertile, well drained soil. The modern varieties which remain open during the day are more ornamental; they are compact enough not to require staking. The Tobacco Plant suffers only from aphids, against which it is possible to spray.

FEATURES

Height: 2 feet (60cm)
Flowering period: June to September
Colour: Greenish yellow
Scent: Very scented at night

1 Stigma
2 Petal
3 Stamen

1 *3* *2*

PAEONIA OFFICINALIS

*L*IKE many of our best-loved plants, Paeony (or Peony) was known to the Ancient Greeks and consequently has a mythological name. It was called after Paeon, a physician who used its roots to cure Pluto when he had been injured by Hercules. As with the Mandrake, it was thought that great care had to be taken to avoid pulling up the roots but, by the time Gerard was writing, this tradition was regarded as a "vaine and frivolous" superstition, "for the roote of Peionie, as also the Mandrake, may be removed at any time of the yeere, day or hower whatsoever". Good old Gerard, striking a blow for modern rationalism! Despite these brave words, however, it would be surprising if he had never fallen prey to superstitions just as "vaine and frivolous".

P. officinalis came to Britain from the Mediterranean region before the middle of the 16th century. Within 20 years, according to William Turner, it had become common everywhere. By 1629, Parkinson was growing the two double varieties as well. (The double red, especially, is still widely cultivated in cottage gardens, despite the gorgeousness of the modern cultivars.) At that time the roots were thought to help in the treatment of epilepsy, and so unfortunate children had them hung around their necks.

This plant grows up to 2 feet (60cm) tall. The leaves, which colour interestingly in autumn, are deeply cut into segments, these segments being as much as 4 inches (10cm) long. The flowers of the type species are blood-red, 5 inches (13cm) across, bowl-shaped, and produced in May and June. These days the doubles, 'Alba-Plena' and 'Rubra-Plena', are much more commonly grown.

CULTIVATION

Paeonies cause consternation: they have an infuriating habit of not flowering for several years after they have been planted. This will not happen, however, if it is appreciated that the crown with the buds must be no more than 1 inch (2.5cm) below the surface of the soil. Generally speaking, it is fair to say that paeonies prefer to be left where they are and not moved about regularly; indeed, they have been known to inhabit the same spot happily for half a century or more.

Paeonies like a well dug fertile soil, and water in dry weather; they also enjoy an annual spring mulch. They will almost certainly lean if not staked, especially if rain falls on the heavy flower-heads of the double forms. They *can* be divided, if necessary, in September, and the species may be grown from seed, sown in that month and put in a cold frame. They can suffer from leaf spot damage and also paeony wilt.

FEATURES

Height: 14-24 inches (36-60cm)
Flowering period: May
Colour: Red

1 Petal
2 Stamen
3 Stigma
4 Ovary

PAPAVER SOMNIFERUM

Opium Poppy

*E*VEN the most determined non-gardener has heard about this plant, because it is the source of opium and, therefore, heroin. At the same time it is an ornamental and garden-worthy plant, provided neither its capacity for prolific seeding nor its unpleasant smell irritate excessively. It is native to the Mediterranean countries and the Middle East, and the Romans probably brought it to Britain as a weed seed in fodder and on the hooves of their horses — which is why it became naturalized along the sides of Roman roads. Parkinson mentions double varieties coming in from Constantinople at the end of the 16th century. This plant was grown in Britain commercially in the early 19th century, both for opiates and for poppy seeds, which contain no narcotic but are rich in an oil as well as being useful for sprinkling on bread. Whether this poppy was grown in Britain to make the narcotic opium or whether it was cultivated only to provide the active ingredient for the "medicinal" laudanum is a moot question. Opium is obtained in the East from the sap of the half-grown seed capsules, but the conditions are not sufficiently hot in Britain for the sap to form properly here.

P. somniferum is an annual species which grows up to 2½ feet (76cm) tall. It has smooth, green, heart-shaped, toothed leaves and mauve, pink, white or red flowers up to 4 inches (10cm) across, borne on long stems. The petals are often wavy and the stamens are yellow. The flowers come out between June and August; after fertilization, flat-topped round seedheads full of seed are produced. (Linnaeus once computed the number of seeds there were to a seedhead; he counted 32,000!) There is a variety called *P. paeoniaef-lorum* which has double flowers; it is so-called because of its resemblance to the common double paeony.

The most frequently planted and useful perennial species is *P. orientale*, from which have come many good named forms over the years. This plant is native to that part of Turkey called Armenia and was grown by George London before 1714. The type plant has scarlet flowers with a black blotch at the base of the petals, but there are many other coloured forms grown, notably 'Mrs Perry' (salmon-pink), 'Perry's White', and 'King George' (scarlet).

CULTIVATION

This plant could not be easier to grow; indeed, growing it is too easy at times, for it can threaten to become a weed in flower borders. It certainly enjoys a sunny position but is otherwise not fussy. The only thing that it does not care for is transplantation. Opium poppies have strong enough stems not to require staking — indeed, the only care needed is the punctilious removal of the dead heads to prevent them from seeding. These poppies should be propagated in spring, or even in September, by sowing the seed in the place where you want the flowers and covering it with a very thin sprinkling of soil. The seedlings can be thinned after they have germinated. Occasionally these plants get downy mildew, but on the whole they enjoy the best, and rudest, of health.

FEATURES

Height: 2½ feet (76cm)
Flowering period: June to August
Colour: Various — mauve, pink, white, red
Scent: Unpleasant

1 Seeds
2 Seed case

PHILADELPHUS CORONARIUS

*T*HERE is an interesting history attached to the Philadelphus. It was introduced to Europe by the Holy Roman Emperor's Ambassador to the Court of Suleiman the Magnificent, Ogier Ghiselin de Busbecq, who brought it back with him on his return to Vienna in 1562. He also brought with him the Common Lilac (*Syringa vulgaris*) at the same time, and the names became confused. That is why, even today, *Philadelphus* is often called *Syringa*, properly the Latin name for Lilac. The wood of both these plants is pithy in the middle, so stems could be made into pipes; consequently these plants were originally called the White Pipe and the Blew Pipe trees.

Philadelphus had arrived in Britain before 1597. Gerard mentions in that year that he has it flowering in his garden. Curiously, he did not care for the fragrance which we find so delicious — indeed, he relates that when flowers were put in his room the smell woke him up and so he "cast them out of my chamber".

Philadelphus means "brotherly love"; the reason for the name is unclear. The common name, Mock Orange, relates to the similarity of the scent to that of orange blossom. *P. coronarius*, the earliest import, has been superseded by newer, better, largerflowered varieties. It is, however, very reliable, floriferous and fragrant. It makes a round, spreading bush up to 9 or 10 feet (2.75-3m) high; the leaves are rather dull, being ovate, toothed and mid-green in colour. The flowers, which appear in June, are not completely white; there is a dash of cream in them. A form of this plant, called 'Aurea', has yellow leaves and is very popular, despite the fact that the leaves burn up if the plant is grown in full sun. There is also an uncommon variegatedleaved variety as well as a dwarf form but, as this latter flowers only sparsely, it hardly seems worth growing. The modern hybrids, which are preferable to *P. coronarius*, have its blood in them. Some of the best are 'Manteau d'Hermine', "Virginal', and 'Belle Etoile'.

CULTIVATION

This plant is extremely easy to cultivate, and usually flowers very well even when thoroughly neglected. It does not really need to be pruned, although thinning out older wood after it has flowered is helpful to it. Apart from blackfly in some years, it is largely immune to pest and disease damage. It will grow in an ordinary (even dry) soil, in either sun or partial shade, although it flowers better in sun. Cuttings taken in July will root in a cold frame, and those hardwood cuttings taken in October can be rooted in a nursery bed outside.

FEATURES

Height: Up to 9-10 feet (2.75-3m)
Flowering period: June
Colour: White
Scent: Overpowering

1 Petal
2 Stigma
3 Stamen

PLATYCODON GRANDIFLORUS

Balloon Flower; Chinese Bell- flower

THIS intriguing plant is closely related to the Campanula, and derives its common name from the shape of the flower-buds, which do resemble inflated balloons. It has always been much prized in its native northern China, Siberia and Japan for its medicinal properties. It arrived in Britain in 1782, but these days is not much grown. Far more common is the compacter variety of it which was sent home to Britain from Japan by Charles Maries. He was a professional collector sent out by the Veitch nursery in 1877, and he found *Platycodon grandiflorus* var. *mariesii* on the island of Yezo. The type species grows up to 2 feet (60cm) tall and has oval leaves and very pretty china-blue, saucer-shaped flowers in succession from June until August. *P.g.* var. *mariesii* grows to only 1 foot (30cm) tall.

CULTIVATION

Platycodons do not present many problems in cultivation, provided you take into account the fact that they are late to start into growth in the spring: it is all too easy to decapitate the new growths with a careless hoe, having quite forgotten that the plant was there at all. Otherwise, they are quite happy to remain undisturbed in a reasonably fertile and drained garden soil in a sunny place. Gertrude Jekyll grew both the usual type and the dwarf *P.g.* var. *mariesii* in her rock garden.

Propagation is best done by seed, for the roots do not like to be dug up and divided. Division should be done in March, if at all. That is also the time when seed is sown, either thinly in drills outside or in a seed tray under glass. The little plants have to be handled very gently because of the incipient roots, and planted out when those roots are dormant, in late autumn or winter.

FEATURES

Height: 2 feet (60cm); 1 foot (30cm) for *P.g.* var. *mariesii*
Flowering period: June to August
Colour: China blue

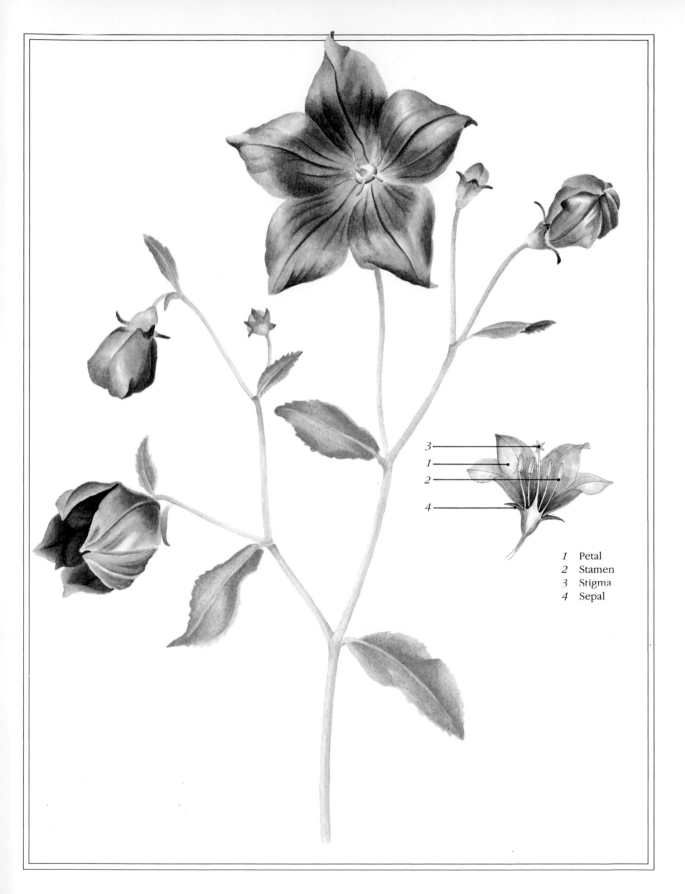

1 Petal
2 Stamen
3 Stigma
4 Sepal

PLATANUS ORIENTALIS

Oriental Plane; Oriental Sycamore

*P*LATANUS ORIENTALIS is famous as much for the fact that it is a parent of the London Plane as for its own qualities. It is, however, a tree well worth growing. It has a long and distinguished documented history, and has been domesticated and planted since ancient times in Europe and northern India (Kashmir). It comes from Greece and other countries which border the Adriatic: Yugoslavia, Bulgaria and Albania. Hippocrates, the "Father of Medicine", taught his students under an Oriental Plane which, supposedly, can still be found on the island of Cos. He lived in the 5th century before Christ, so this story rather strains one's credulity. In Turkey there still exists a plane near which crusaders are supposed to have camped during the First Crusade. This tree has several stems which are fused together, a fact which foxed the French botanist Augustin Pyrame de Candolle, who estimated in the early 19th century that it was 2,000 years old. It is probably much younger than that.

The Oriental Plane was first brought to Britain at the end of the 16th century, probably from the Near East by an agent of the Levant Company. William Turner, in 1548, maintained that he knew two specimens of it in England, but it seems likely that he misidentified it (or was mistaken). *P. orientalis* is not planted very much these days, having been thoroughly supplanted by its offspring,

Plantanus × acerifolia, the London Plane, but there are some very good old specimens, such as the one at Rycote Park, Oxfordshire.

P. orientalis differs from the London Plane in having a shorter and more rugged trunk and leaves which are more deeply lobed. It has a very large and handsome spreading crown of branches. The leaves are up to 9 inches (23cm) wide, with five or seven deep lobes, and the fruits are balls, two to six of which hang on a long stalk. The London Plane is a hybrid between *P. orientalis* and the Western Plane, *P. occidentalis*; it is supposed to have been raised before 1700, perhaps near Oxford.

CULTIVATION

The Oriental Plane is very easy to grow; not only will it grow happily in any reasonable soil provided the position is not shaded, but it is tolerant of hard pruning if planted in too confined a space. Cuttings are quite successful if taken in October and put in a sheltered place outside. This tree is usually healthy and tolerant of smoky atmospheres.

FEATURES

Height: Up to 100 feet (30m)
Flowering period: April to May; the fruits hang all winter
Colour: Pale green

1 Pericarp
2 Seed coat
3 Endosperm

PRIMULA AURICULA HYBRIDS

Dusty Miller' Bear's Ears;
Mountain Cowslip

∽

*F*ROM the middle of the 16th century onwards, Protestant refugees
known as Huguenots fled from perse-
cution in Catholic France to find a
haven in Protestant England. They
brought with them, as well as their reli-
gion and their artisan's skills, living
specimens of *Primula auricula*, an
Alpine primula long in cultivation in
Europe. In the 1570s, a hybrid between
P. auricula and *P. rubra* called *Primula*
× *pubescens* was found in the Royal
Gardens in Vienna, and sent by Clusius
northwards in 1578. *P.* × *pubescens* was
probably the parent of the Alpine Auriculas, whereas *P. auricula* was the fore-
runner of the Show Auriculas. Both
Gerard and Parkinson knew auriculas,
the latter commenting that the flowers,

> *being many set together upon*
> *a stalke, doe seeme every one of*
> *them to bee a Nosegay alone of it*
> *self; and besides the many*
> *differing colours that are to be*
> *seene in them, as white, yellow,*
> *blush, purple, red, tawney,*
> *murrey, haire colour ... which*
> *encrease much delight in all sorts*
> *of the Gentry of the Land, they are*
> *not unfurnished with a pretty*
> *sweete sent, which doth add an*
> *encrease of pleasure in those that*
> *make them an ornament for their*
> *wearing.*

In 1665, Rea described four classes of
auriculas, and it is clear that many kinds
were being cultivated, including striped
varieties which are no longer grown. By
the middle of the 18th century cultivated auriculas had edges to the petals,
and were grown as florist's flowers,
usually in pots under glass, for,
although perfectly hardy, the flowers
were easily damaged by bad weather in
the spring.

While the Paisley weavers were
busy with their pinks during the early
years of the 19th century, the silk
weavers of Lancashire and Cheshire
turned their energies towards cultivating auriculas. It is scarcely a coincid-
ence that many of the Huguenot
refugees settled in the northwest of
England and continued their old craft
of weaving. It is said that, at that time, a
Lancashire workman would pay two
guineas for a new variety, although at
the very most he earned 30 shillings a
week. Many shows were held, and in
1872 the National Auricula Society was
founded.

The Show Auricula invariably has
smooth leaves, often dusted with the
floury "meal" so common amongst primulas. (The Alpine Auriculas have no
"farina".) These grey-green, ovate
leaves give point to the name of ' Bear's
Ears'. The flowers, about 6 inches
(15cm) tall, are held in clusters above
the leaves and have an "eye" of a different colour from the rest of the flower.

Growing them for show in the
1800s must have been a nightmare: the
flower had to be one and nine-sixteenths of an inch (4cm) across, the
"eye" nine-sixteenths of an inch
(1.4cm) across, and the tube one-six-
teenth of an inch (1.6mm) across. The
body colour of the flower had to be
black, and the "eye" had to be covered
with a farina. No wonder the craze for
growing them for show has waned!

CULTIVATION

Alpine Auriculas, of which 'Yellow
Dusty Miller' is the best known of those
that remain, are happiest in a fertile but
well drained gritty soil — after all, they
are alpine plants — but are not fussy
about whether they grow in full sun or

1 Petal
2 Stigma
3 Stamen
4 Ovary

in semi-shade. They can also be grown in a gritty compost in a pot, and staged in a cold house or alpine house; this has the advantage of protecting the flowers while they bloom from April until June.

Propagation consists of detaching any side shoots with roots, at any time between June and August, and planting them in a cold frame until they are firmly established. It is also possible, although technically more difficult, to take small cuttings at this time. Seedsmen do sell auriculas as mixtures, and these can be sown either in heat under glass or in trays in a cold frame, provided the seedlings are shaded from the sun. Once well germinated, they should be potted up in small pots and plunged outside until large enough to transplant to flowering positions or bring into the glasshouse to flower.

FEATURES

Height: Up to 9 inches (23cm)
Flowering period: April to June
Colour: Myriad; "eye" different
Scent: Strongly of honey

PRIMULA FLORINDAE

Giant Cowslip

*I*N *Pilgrimage for Plants* (1960), Frank Kingdon-Ward describes a plant which he discovered in southeast Tibet in 1924 and named after his first wife, Florence.

The most frequent adverse criticism of Primula florindae *is that it is rather coarse, by which the critic invariably means large. Certainly it* is *large, much larger than most* [Primula] *species met with in gardens. But no one who had seen it growing by the great river of Tibet would have thought it large by Tibetan standards. It is a matter of scale, and of proportion. Grown in a small mound built of small, haphazard bits of rock* [i.e., *the average British rock-garden*], *among which small plants are perched like flies on a stale bun,* P. florindae *can look as coarse as Gulliver looked among the Lilliputians; a triton among minnows.*

He goes on to say:

A single plant is big and bold enough to stand alone if the surroundings are big and bold. But P. florindae *is one of those plants which looks at its best growing in a crowd, as nature intended it to grow, challenging the current. It is the forest of stems rising above the curved leaves, like frail masts from a rough green sea, which is the attraction; not the lone specimen ... The size of the large basal leaves ensures room for the flower scapes to display themselves; but the flowers themselves are so crowded that while they shake themselves free from the mop you get the feeling the remainder must explode like a rocket, and send a cloud of scented yellow stars drifting to earth.*

P. florindae grows up to 3 feet (90cm) tall, which is very big for a primula. It has ovate, toothed leaves which will attain 8 inches (20cm) in length, and the drooping bell-shaped flowers are borne in umbels at the top of the long, stout stems in June and July. They can vary in colour from yellow to orange and even to red, but are most usually a lemony yellow.

CULTIVATION

As we know that *P. florindae* has large leaves and grows by the sides of Tibetan rivers, we can safely deduce that it will do best in a moist soil which does not dry out in summer. It should be planted in suitable conditions any time during the dormant season, in either full sun or semi-shade. Peat and other organic matter should be dug into the soil before planting, and used as a mulch as well. This plant is easily divided in July after it has finished flowering. Ripe seed should be sown in trays and put in a shaded cold frame. The seedlings can be planted out in nursery rows until large enough for transplanting to flowering positions. Altogether, it is one of the easier Asiatic primulas to deal with.

FEATURES

Height: Up to 3 feet (90cm)
Flowering period: June and July
Colour: Lemon yellow
Scent: Yes

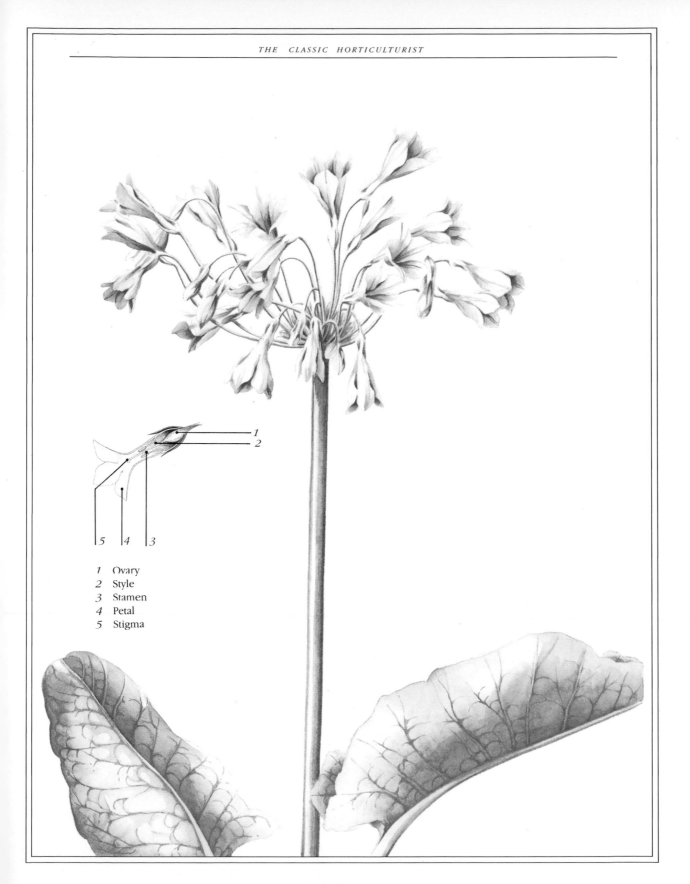

1 Ovary
2 Style
3 Stamen
4 Petal
5 Stigma

PRUNUS 'TAI-HAKU'

Great White Cherry

*P*RUNUS 'Tai-Haku', the Great White Cherry, is one of the 'Sato Zakura'; that is, the Japanese garden cherries. Like the others, it probably descends from the Hill Cherry, *Prunus serrulata* var. *spontanea*, but there may be other influences at work as well. It was grown in Japan for a long time, chiefly around Kyoto, but died out from there early in the 20th century. In 1923, Captain Collingwood Ingram, the great British authority on Japanese cherries (so much so that his nickname was "Cherry" Ingram) went to see a Mrs Freeman and her garden in Sussex. She told him how, in 1899 during a visit to Provence, she had met a Frenchman who had a Japanese friend prepared to send Japanese cherries to England. Mrs Freeman wrote to the Japanese friend and in the spring of 1900 received a small collection of cherries from him. Some of these plants had never been seen in England before. In Mrs Freeman's garden Ingram found one which he failed to recognize. It was almost dead, but he managed to propagate it from a few pieces of budwood, and discovered that this was 'Tai-Haku', a cherry which was by now extinct in Japan. Soon afterwards he had the satisfaction of reintroducing the plant to that country.

'Tai-Haku' is a very vigorous tree, and the flowers, which are large — 2½ inches (6cm) across — are of the purest white and in April make a stunning contrast to the leaves. These are bronze-red as they unfold, only later turning to green; in autumn they turn red and yellow before they fall. 'Tai-Haku' has a distinctive bark with very big air-pores

in it. This tree can grow to 25-30 feet (7.6-9m) high and spread a similar amount. It well deserves the title 'Great White Cherry'.

CULTIVATION

'Tai-Haku' is not a difficult ornamental cherry to grow, provided its considerable spread is taken into account so that pruning is not necessary. The soil into which it is planted should not be too dry or, conversely, waterlogged; the soil nearby should not be deeply cultivated after planting, because the tree's roots tend to be shallow. The Japanese cherries are usually grafted plants and therefore not easy to increase. The best method is to bud the scion-wood onto a *P. avium* stock.

This cherry is, like the others, very popular with bullfinches in winter, especially in country districts, so netting is advisable where possible. Aphids are a nuisance, as are caterpillars, so a tar-oil winter wash is helpful. Ornamental cherries naturally suffer from the same diseases which afflict the fruit trees; for example bacterial canker, silver leaf and fireblight. Attack by the first necessitates the cutting off of the cankered wood; attack by the last two requires at least the pruning of affected branches, and often the grubbing up and burning of the whole tree.

FEATURES

Height: Up to 30 feet (9m)
Flowering period: April
Colour: White; young leaves bronze

1 Style
2 Petal
3 Stamen
4 Ovary

PUNICA GRANATUM

Pomegranate

*T*HIS plant has been the subject of many myths over the centuries, perhaps the most famous being that of the ill fated Persephone, who was condemned to spend four months a year in the Underworld because she ate half a pomegranate seed given to her by Hades.

The Pomegranate is not a common shrub in Britain mainly because it is not very hardy. Its Latin name is *Punica granatum*; the Romans called it *Malus punicum*, because they obtained it from the area around Carthage, or *Malum granatum*, because of the grain-like seeds that the fruit contains. It receives a mention in the Bible, in *The Song of Solomon*. It originally came from Iran and Afghanistan, but is now grown all round the Mediterranean; it has been cultivated in Britain since the 16th century. It is, unfortunately, disappointing in Britain, because the summers are only rarely hot enough to ripen the fruit properly. However, it does have very striking bright scarlet flowers from June until September. Philip Miller reported that the fruits rarely ripened sufficiently outdoors to be eaten, but nevertheless "they made a very handsome Appearance upon the Trees". According to Parkinson, the fruits could be used to make a permanent ink.

P. granatum is a deciduous shrub. The flowers are tubular with five to seven petals; the leaves are ovate or oblong and shiny. The fruit is a berry with a leathery skin and soft flesh which is formed by the outer seedcoats (arils) of the numerous seeds. The juice from the fruits can be used to make the drink Grenadine.

CULTIVATION

Pomegranates are not hardy in Britain anywhere except in the southwest, and even there they will not produce the fruits for which the plant is famous. *P. granatum* is best grown, therefore, in a greenhouse or conservatory, at a minimum winter temperature of 5°C (41°F). For good fruit formation, an autumn minimum temperature of 13°C (55°F) is necessary. It likes a well ventilated atmosphere in all but the coldest weather and a regular feed in the summer. Outdoors, it does best if planted in late spring against a warm wall; in these conditions it will in time make a large shrub.

The quickest way of acquiring new plants is to take cuttings of lateral shoots (with a heel) in late July and strike them in a heated propagator. The rooted cuttings will need to be grown on in a cool greenhouse until large enough to plant out. Layering is suitable for outdoor specimens.

FEATURES

Height: Up to 10 feet (3m)
Flowering period: June to September
Colour: Scarlet; fruit a deep yellow or orange

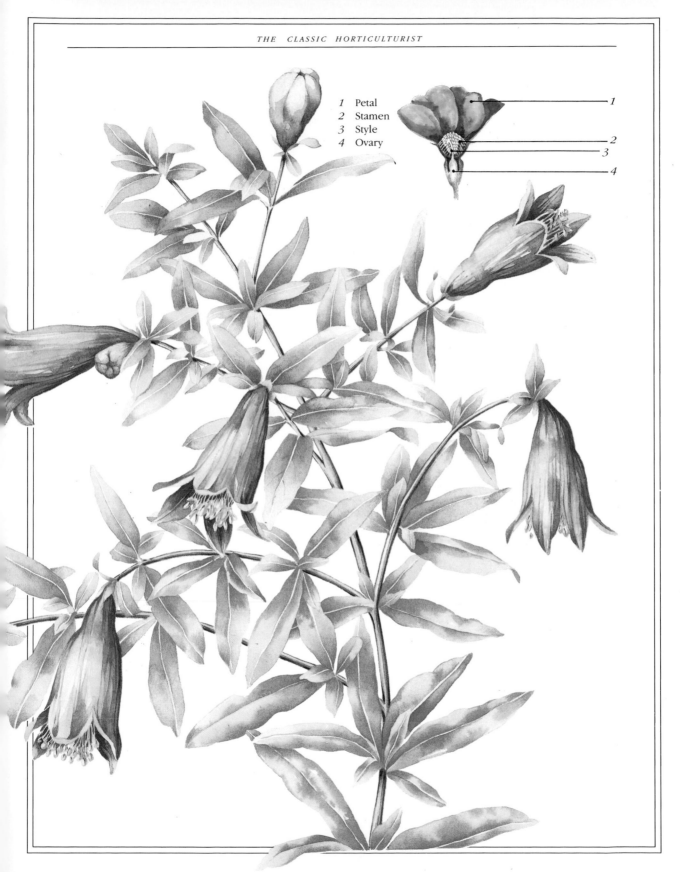

1 Petal
2 Stamen
3 Style
4 Ovary

1
2
3
4

RHODODENDRON ARBOREUM

The Tree Rhododendron

*R*HODODENDRON, meaning a rose tree, is the name given to this genus by Linnaeus. *Rhododendron arboreum* was the first Himalayan species to be collected and introduced to Europe. A red-flowered form was first found, by a Captain Thomas Hardwicke of the Indian Army in 1796: he found it flowering in Kumaon, southeast of Dehra Dun, in the Himalayas. However, this Rhododendron is reputed not to have flowered in England until 1825, so Hardwicke's original must have been lost and a new introduction made at some time in the first few years of the 19th century. Either that, or the plant was not sent home at the time of discovery.

R. arboreum is tender, except in the south and west of England. It does well in Cornwall, especially, where it attains almost tree-like proportions, as befits its name. Its children make up a sizeable proportion of our modern hardy hybrid garden rhododendrons. One of the earliest hybrids was raised in 1826 at Highclere, the seat of the Earl of Caernarvon, and was called *R.* × *altaclarense*. It was a cross between *R. arboreum* and a *R. catawbiense/R. ponticum* cross. This hybrid marked the beginning of the raising of our modern garden rhododendrons.

In a good situation this evergreen rhododendron can grow up to 30 or 40 feet (9-12m). It has oblong-lanceolate and very leathery leaves (with a brown felt underside) up to 8 inches (20cm) long. The flowers, which are borne in dense trusses of about 20, normally bloom in March or April. Besides the hybrids, there are also several forms of the species, including *R.a. cinnamomeum* (usually white-flowered and with cinnamon-coloured undersides to the leaves), which is the hardiest, and *R.a. campbelliae*, which has purple-pink flowers.

CULTIVATION

This tree will never be widely planted because it is not very hardy and, even in the south of England, flowers too early in the season for its own good; the flowers are often frosted. It is, therefore, suitable only for Cornish, Devonian and western Scottish gardens where spring frosts are rare. Also, it does not flower for at least 10 years after planting, and eventually grows so large that it is suitable only for extensive gardens. Some of its hybrid offspring, however, such as the wonderful 'Pink Pearl', are, although large, more rewarding to cultivate. Rhododendrons in general like cool, moist, semi-shaded, sheltered conditions in enriched medium soil; with the exception of *R. hirsutum* and 'Cunningham's White', they will not tolerate any lime in the soil. If the hardy hybrids are grown in sunny places, they have to be mulched annually so that their roots do not dry out.

Propagation of species rhododendrons is by sowing seed in early spring in heat, or later in the spring in a cold frame. When they have germinated, they should be potted up in a leafy compost until large enough (the following spring) to be planted out in nursery rows. Layering is slow, but successful for rhododendrons with branches near the ground. A slit is made on the underside of the branch which is then pegged to the ground and buried by soil. The tip is staked in an upright position. After about 18 months or two years, when the layer has made roots, it can be cut from the mother-plant and transplanted.

Height: Up to 40 feet (12m)
Flowering period: March and
April
Colour: Deep red

1 Petal
2 Style
3 Stamen
4 Ovary

'ROSA MUNDI'

(R.gallica 'Versicolor')

ONE might be forgiven for thinking that a rose with a name like 'Rosa Mundi' must be a species, but it is only a form of *Rosa gallica*; its other, less common, name is *R. gallica* 'Versicolor'. It is a semi-double sport of the so-called 'Apothecary's Rose' known also as 'Officinalis' or 'The Red Rose of Lancaster'. This has led people, erroneously, to call 'Rosa Mundi' the York and Lancaster Rose, the honour of which must properly go to a cross between 'Officinalis' and *Rosa × alba semi-plena*.) 'Rosa Mundi' is supposed to be named after the mistress of Henry II called Fair Rosamund, Rosamund Clifford. On her tombstone in the nunnery at Godstow Priory, near Oxford, were the words *Hic jacet in tomba rosa mundi, non rosa munda.*

'Rosa Mundi' makes a prickly lax shrub up to 4 feet (1.2m) tall. The three to five leaflets per petiole are oval, rough and leathery. The June-flowering blooms are semi-double and loose-petalled and, in colour, pinkish-white streaked and striped with purple, bright red and pink. If grown on its own roots, the plant suckers freely, and this leads, in time, to the development of a rose thicket. As it is a sport of 'Officinalis', one of its branches will sometimes revert to the colour of that rose.

CULTIVATION

'Rosa Mundi' is often to be seen as a low, lax hedge (something Vita Sackville-West strongly recommended) and is a marvellous sight as one in June, especially if associated with the darker shades of Sweet William. However, its foliage is not sufficiently attractive for one to wish to see it for the rest of the season, when the rose is out of flower, so it is unwise to plant 'Rosa Mundi' too prominently. It is also famously prone to mildew. The mildew must be tackled with regular spraying with a fungicide in the summer months. If grown as a hedge (or as a specimen, for that matter) 'Rosa Mundi' will benefit from being clipped over in February; more sophisticated pruning is not necessary. Gallica roses require a less rich soil than many roses, but they appreciate a mulch in the spring to conserve moisture and add some nutrient to replace what they and the rain have removed in the previous season.

FEATURES

Height: 3-4 feet (90-120cm)
Flowering period: June
Colour: Pinkish-white, streaked with bright red
Scent: Sweet

1 Petal
2 Stamen
3 Stigma
4 Ovary

SALVIA PATENS

*L*ITTLE is known about the history of this plant except that it was introduced to Britain from Mexico in 1838 and that seed was sent back on occasion by travellers who were interested amateurs attracted by the glorious colour of the flowers as well as by professional collectors. Robinson said that its brilliance was equalled by few flowers in cultivation. Gertrude Jekyll described its colour, along with that of Morning Glory and some of the gentians, as "perfectly pure blues. They are none too many and are, therefore, all the more precious in garden use." She used *Salvia patens* for summer border schemes, despite the fact that it was not hardy, by planting pot-grown plants in late spring. It is a perennial (in warm places) which grows no more than 30 inches (76cm) tall. The stem is hairy and, like all labiates, square in section. The leaves are oval and pointed, and the flowers, in colour the purest and deepest sky blue, are borne in August and September. There is a paler form called 'Cambridge Blue' as well as a white form called *S.p.* var. *alba*.

CULTIVATION

The difficulty experienced by many gardeners in hanging onto this plant for long arises because of its tenderness. It will survive a mild winter in most districts, provided it is planted deeply and protected, but many people find it easier to treat it as a half-hardy annual and resow the seed each year. In that case this plant is a candidate for the standard half-hardy annual treatment: sow the seed in heat in March, prick out into trays, harden off the plants in a cold frame, and plant out in mid- to late May. Salvias like a rich but light soil.

Cuttings can also be taken in September, and rooted in a propagating frame. The rooted cuttings can be planted out in May, after being hardened off, and they will flower earlier than seed-raised plants.

FEATURES

Height: 30 inches (76cm)
Flowering period: September
Colour: Sky blue
Scent: Aromatic leaves

1 Stamen
2 Style
3 Petal
4 Sepal
5 Ovary

SCHIZOSTYLIS COCCINEA

Kaffir Lily; Crimson Flag

S CHIZOSTYLIS COCCINEA, known as the Kaffir Lily, comes, like so many good "bulbs", from South Africa. It was introduced to Britain in 1864. In October and November it has rich red star-shaped flowers in spikes on long — up to 3 feet (90cm) — stems above grassy, sword-shaped leaves. The unusual time of flowering, coupled with the innate charm of the flowers, makes this a highly desirable plant, and its variety 'Major' is worth seeking out for its larger flowers set on even sturdier stems. The most vigorous is 'Viscountess Byng'; this is pale pink and blooms late, in November. (Viscountess Byng herself is supposed to have been unable to grow this plant successfully in the dry soil of Essex, which must have been something of a disappointment to her.)

In 1920 an English nurseryman holidaying in the West of Ireland came across a schizostylis with clear pink flowers in the garden of a village doctor. He bought the entire clump for £50 and showed the plant the following year in London under the name of 'Mrs Blanche Hegarty', thus ensuring immortality for the doctor's wife.

CULTIVATION

Schizostylis are South African rhizomatous perennials but nevertheless require a moist soil which will not dry out in summer. That means planting them in a sunny place which is, however, damp (easier said than done) and mulching them in spring with peat or similar material. Watering in dry summers is a help, as is, in cold winters, protecting their roots in exposed places in winter with bracken or straw. *Schizostylis* can be grown as pot-plants in conservatories, being plunged out of doors for the summer, regularly fed to help flower-bud initiation, and brought inside once they start to flower. If the plant is happy in the garden and thrives, it will need dividing every three years, as the clumps can become very congested. This entails pulling the clumps apart in the early spring and replanting them in groups of about five shoots. The main enemy of *Schizostylis* is *Botrytis*, which can affect them in the autumn; spraying with a fungicide will help.

FEATURES

Height: 2½-3 feet (76-90cm)
Flowering period: October and November
Colour: Scarlet

1 Style
2 Petal
3 Stamen
4 Ovary

SORBUS 'JOSEPH ROCK'

Joseph Rock's Rowan

THE exact provenance of this plant is something of a mystery. The original plant is at the Royal Horticultural Society's Garden at Wisley, where it has grown since seed of it was sent from the Royal Botanic Gardens, Edinburgh, in 1937, under a collecting number belonging to Dr Joseph Rock, a US collector. Rock had found this tree in Yunnan, China, in 1932 and sent seed to Edinburgh. However, the Wisley tree differs from the one grown under that number at Edinburgh, and why this should be so has never been satisfactorily explained. So the name applies only to those plants propagated from the Wisley plant, and its parents (for it is a hybrid) cannot be established beyond doubt. However, despite its mysterious origins, this very fine tree has become deservedly popular. Rock was, like many of the plant collectors, slightly larger than life. Although a naturalized US citizen, he was of Austrian descent and completely self-taught (he learned Chinese at the age of 13). He was sent to Burma to discover and collect seed of the Chaulmoogra tree, whose oil is known to be helpful in the treatment of leprosy, and he remained for nearly 30 years before World War II collecting plants and studying primitive peoples in China. His most famous introduction is *Paeonia suffruticosa* 'Rock's Variety'.

Sorbus 'Joseph Rock' makes a small tree, no more than 30 feet (9m) tall, with branches which are upright to begin with but which spread out with age. The leaves, which can be as much as 8 inches (20cm) long, have between 15 and 19 leaflets. Each leaflet is oblong and sharply toothed. The leaves colour in the autumn to a mixture of red, purple and orange. The flowers are cream and carried in May. The berries, which are round and borne in clusters, start off green, turn to white, and finish a lovely yellow — most unusual for a rowan.

CULTIVATION

Rowans, even ones from China, are not difficult to grow. They will manage well in either a sunny or a partially shaded place but, as their autumn colour is their crowning glory and their leaves colour better when they have had full sun upon them, it seems only sensible to grow this plant in a sheltered sunny position if at all possible. Rowans are rarely troubled by pests but they have the weakness of all Rosaceae for fireblight and canker, and will sometimes succumb to silver leaf as well. Fireblight is a notifiable disease because of the havoc it can wreak in commercial orchards, and a bad attack usually warrants destruction of the plant; canker can, and should be, cut out.

Branches affected by silver leaf should be removed and, if fungi appear on the trunk, the tree should be grubbed.

FEATURES

Height: Up to 30 feet (9m)
Flowering period: May; fruits in August and September
Colour: Cream flowers; yellow fruits; scarlet, crimson and purple autumn colour

1 Style
2 Stamen
3 Petal
4 Ovary

SYRINGA PERSICA

Persian Lilac

SYRINGA PERSICA was brought to Europe, like the Common Lilac, thanks to the good offices of a diplomat — in this case the Venetian ambassador to Constantinople, who brought it back with him some time before 1614. It is supposed to have been introduced to Britain by John Tradescant the Elder about 1620, after he had volunteered to join the pinnace *Mercury* to fight against the corsairs of the Barbary Coast who were giving trouble to British shipping in the Mediterranean. That may or may not be true, but certainly by 1640 Parkinson was reporting that Tradescant was growing it at Lambeth. For a long time the plant was thought to be a jasmine — indeed, Hanmer referred to it as such in 1659. The word "syringa" comes from the Greek *syrinx*, meaning "panpipes", and alludes, as we have noted before, to the fact that the stems, if hollowed out, can be used to make musical pipes. By 1785, three forms were being grown: a blue one, a white one and a cut-leaved variety. About 1777, in the Botanic Gardens at Rouen, the ordinary Common Lilac was crossed with it, the result being the hybrid called the Rouen lilac (*S. × chinensis*). This had arrived in England by 1795.

S. *persica* is a deciduous shrub, growing to no more than 7 feet (2m) high; it is, therefore, more compact and less straggly than the Common Lilac. It is bushy and rounded, with lanceolate leaves that are green and up to 2½ inches (6cm) long and half an inch (1cm) wide. It has flowers which are pale lilac, and it is very scented. The flowers come out in May in short 3-inch (7.5cm) panicles at the top of last year's growths. There is a white form.

CULTIVATION

Provided syringas are planted in a reasonably fertile soil where some sun can reach them, they are no trouble to grow at all. Nor are they difficult to propagate if semi-ripe cuttings are taken in late summer and put in a propagating frame. After they have rooted, they can be grown on in pots in a cold frame and then, perhaps, planted in a nursery row until large enough to plant out safely. They need little pruning beyond deadheading and the removal of dead and very weak wood at some point after flowering.

FEATURES

Height: 7 feet (2m)
Flowering period: May
Colour: Lilac
Scent: Very scented

1 Petal
2 Stamen
3 Style
4 Ovary

TULIPA CLUSIANA

The Lady Tulip

*T*ULIPS were grown and developed in Turkish gardens by the 16th century and certainly caused excitement in the breast of Ogier Ghiselin de Busbecq, the Emperor Ferdinand's ambassador to Turkey. (He introduced *Philadelphus* to Europe at the same time — see page 124.) Unfortunately, de Busbecq misheard the name of this plant, thinking that the Turks called it *dulban*, which means "turban", when in fact they called it *lalé*. The mistake is odd because "tulip", for that matter does not sound particularly like *dulban* either, and nor does the flower have more than a passing resemblance to a turban. Such considerations matter little; all that is important is that de Busbecq brought it to Europe in about 1554, and Clusius took bulbs to Holland with him when he became Professor of Botany at Leyden in 1593. (There is some doubt, however, that Clusius was the first to introduce the bulbs to Holland, a doubt reinforced by the fact that tulips had already arrived in England by 1578.) There is no space here to recount the tale of "Tulipomania", but we can note that fortunes were won (or, more often, lost) on this most dicey of gambles: the quality of the "breaks" on the flower of the new tulip. It was not until the 1920s that it was properly understood that the streaking on tulips, which made them so desirable and expensive, was the result of a highly capricious virus and ultimately harmful to the plant. By that time the whirlwind of "tulipomania" had long since blown itself out.

T. clusiana, like most of the other species tulips, comes from the hot regions of the eastern and northern Mediterranean — from Iran, Iraq and Afghanistan. It has been grown in Britain at least since 1636. It has solitary flowers which open out into a flat star-shape; these are white inside with a purple blotch in the centre, and white with red stripes on the outside of the petals. The flower is about 2 inches (5cm) in length, and each petal is pointed and something under an inch (1.3cm) wide. Its flowering period is late April and early May. The leaves are very narrow, upright, and blue-green. The whole plant is no more than 9 or 10 inches (23-25cm) high. Yet, despite its diminutive size, it has a charm not vouchsafed to the larger, coarser garden hybrids. For one thing, the stiffness of the stem is not so obvious, and nor is the head out of proportion with the leaves and stem.

CULTIVATION

Because of its places of origin, *T. clusiana* is accustomed to extremes in temperatures and, in particular, to very hot summers. It should therefore be planted in a hot, dry place on the rock garden or in a raised bed, rather than in a conventional border where, unless grown in quantity, it will anyway look rather lost. These tulips also do well in a well drained compost in a container. The bulbs should be planted 6 inches (15cm) deep in November and watered well initially. Unlike many varieties of tulip, which have to be dug up each year because their flowering can begin to degenerate, these can be left alone to multiply if they will. It usually takes five years at least before a species will flower from seed, so it is much more fruitful to dig up the bulbs, remove the offsets, store them in a warm dry place until autumn, and replant them — the smallest in nursery rows and the largest in their flowering positions. Slugs are extremely partial to tulip species, so slug pellets have to be put down. Even worse are the viruses and diseases which can afflict them, especially if they are planted year after year in the same place, although this is admittedly more of a problem with the garden varieties. In general terms, the species tulips are reasonably healthy.

FEATURES

Height: 8-10 inches (20-25cm)
Flowering period: April and May
Colour: White with red stripes on the outside

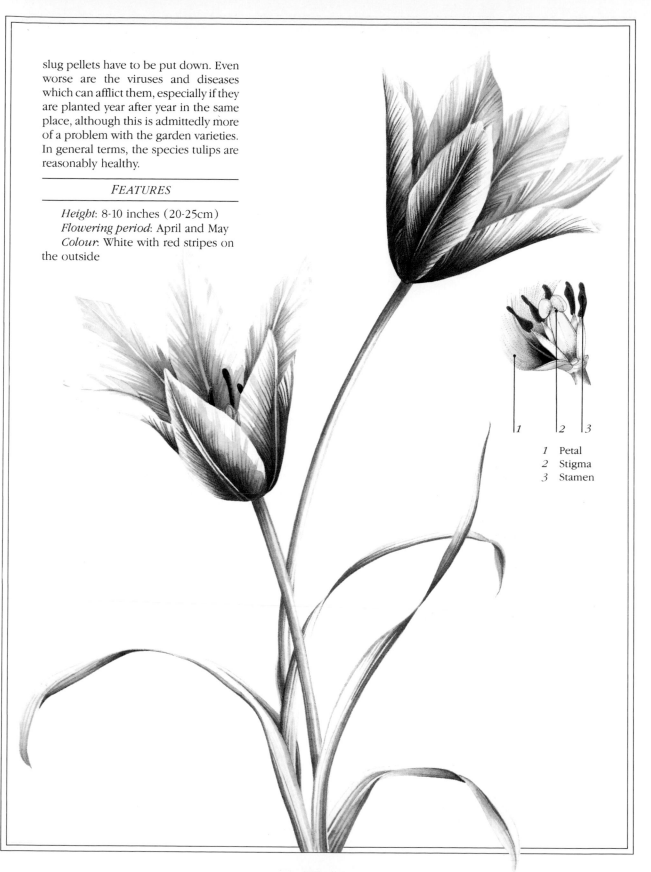

1   Petal
2   Stigma
3   Stamen

VIBURNUM TINUS

Laurustinus; Wild Bay

*T*HIS is a shrub which has been cultivated in Britain for centuries, having been introduced from the Mediterranean region in the 1500s. It was called "Laurustinus" at that time because it was believed to be a kind of laurel, and the name has stuck. It has proved itself always a most useful shrub, not only because of its immense hardiness and its imperviousness to pest and disease attack, as well as to smoky atmospheres, but also because it is winter-flowering — indeed, it is one of the very few genuinely winter-flowering and bone-hardy shrubs the British have at their disposal. It is not everyone's favourite, however. For example, the leaves smell unpleasantly; for E.A. Bowles, who was gifted with an extraordinary olfactory sensitivity, this plant was painful. He wrote of it that "in showery weather [they] exhale an odour of dirty dog-kennel and an even dirtier dog".

V. tinus is an evergreen shrub which makes a dense, round shape, up to about 10 feet (3m) high in time. The branches come down to the ground, which makes it excellent for ground cover. The leaves are glossy green and ovate, and have a prominent central vein. The small flowers, pink-tinged in bud and then white, are borne in terminal clusters from December until April. The fruits are oval and pointed, and coloured the deepest blue. There is a more compact pink-flowered form called ' Eve Price '.

CULTIVATION

This is the viburnum most tolerant of ill-treatment. It thrives best in a sunny position and in good soil, but often it has to make do with less. If it is to give of its best, however, it should be planted in moist rich soil. It can be planted in September or in March/April. It does not require much pruning: a few very old and weak shoots can be cut out after flowering in May. Cuttings root quite readily if taken with a heel in September and put in cuttings compost in a propagating frame or cold frame. This plant can be layered. *V. tinus* has leathery, evergreen leaves which are not much attacked, except occasionally by whitefly.

FEATURES

Height: Up to 10 feet (3m)
Flowering period: December to April
Colour: Pink in bud, white when open

1 Petal
2 Stamen
3 Stigma
4 Ovary

VIOLA TRICOLOR HYBRIDS

Garden Pansies

THE modern Pansy (the name is probably a corruption of the French word *pensée*) is a development of crosses between *Viola tricolor* (Heartsease) and the yellow-flowered *V. lutea*. The pioneering breeding work was done by Lord Gambier's head gardener, Thompson, at Iver in Buckinghamshire in the early years of the 19th century. From 1816 onwards he raised many varieties, but it was not until 1830 that he found one with a central blotch to the flower. This became the ideal for the Show Pansy: a perfectly circular flower with a small blotch in the middle. The rules of the early Pansy Shows, like those of the Auricula Shows at much the same time, were very strict. About 1860 the Fancy Pansies were developed, mainly from pansies bred in France and Belgium; these had stronger colours and a very large blotch on the lower petals. By 1871 these were being shown, and they gradually ousted the Show Pansy as the favourite of enthusiasts.

At that time, pansies were becoming desirable as bedding plants as well. In the 1860s a Scottish nurseryman called James Grieve (probably better known for "his" apple) raised a strain which were crosses between Show Pansies and *V. cornuta*. These became known as 'Tufted Pansies', because their habit was more compact, or 'Violas' — incorrectly, because of course all pansies are Violas. The Latin name of the 'Tufted Pansy' is *V. × williamsii*, to distinguish it from the Garden Pansy, which is *V. × wittrockiana*, although they are often these days all lumped under the heading *V. × wittrockiana*. Examples of *V. × williamsii* include 'Irish Molly' and 'Maggie Mott'. William Robinson was a great devotee of 'Tufted Pansies' because

they were hardy and free-flowering, and he grew them as ground cover beneath his Tea Roses. The 'Tufted Pansy' had a smaller flower than the Show Pansy, and it had a light centre and darker edging.

During the 1880s a Dr Stuart of Edinburgh raised the 'Violettas'. These were smaller than the 'Violas' and had no "rays" to guide insects to the nectaries; they were also scented.

These days the Garden Pansy, *V. × wittrockiana*, which can, with care and the correct choice of varieties, be kept flowering in the garden for 12 months of the year, is by far the most widely grown.

CULTIVATION

Garden pansies are easy enough to plant in either the autumn or the spring in ordinary garden soil in sun or semi-shade. They do best, however, in a fertile soil which is moist but never waterlogged; digging in organic material into a light soil is therefore an advantage. This is a counsel of perfection, it must be said, for pansies will grow, even if not enthusiastically, in any ordinary soil. They are prey to aphids in some years, and their petals often make a delicious breakfast for slugs in the winter, when some varieties resolutely refuse to stop flowering. They do get a disease called pansy sickness, which causes them to collapse. If it occurs, the affected plants are best burned and the soil declared a no-go area for pansies for a while.

Seed of these plants is best sown in July in a seed tray, and put in a shaded place to germinate; the seedlings should be potted on and overwintered in the frame. In favoured localities, pansies can be sown in a row outside in the summer, and planted out in the autumn into flowering sites. Alternatively, an autumn sowing will bring a spring flowering. Cuttings from non-flowering shoots can be taken in summer and put in a shaded frame; this is the only way of propagating the named forms because they may not come true from seed.

FEATURES

Height: 4-9 inches (10-23cm)
Flowering period: Summer-flowering, May to September; winter-flowering, all through mild winters
Colour: Various
Scent: Some

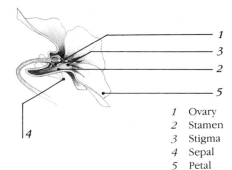

1   Ovary
2   Stamen
3   Stigma
4   Sepal
5   Petal

FOOTNOTES

1. HYAMS, E.   *A History of Gardens and Gardening* Dent, 1971

2. GUNTHER, R. T.   *Dioscorides De Materia Medica* Oxford University Press, 1921

3. HADFIELD, M.   *A History of British Gardening* John Murray, 1960

4. FLEMING, L. & GORE, A.   *The English Garden* Michael Joseph, 1979

5. DESMOND, R.   *Dictionary of British & Irish Botanists & Horticulturists* Taylor & Francis, 1977

6. PARKINSON, J.   *Paradisi in Sole Paradisus Terrestris* Facsimile ed. Dover Publishing Inc, 1976

7. GUNTHER, R. T.   *Early British Botanists* Oxford University, 1922

8. WOODWARD, M. (Ed.) *Gerard's Herbal (1636 Edition)* Gerald Howe, 1927

9. ALLEN, MEA   *The Tradescants* Michael Joseph, 1964

10. LEITH, ROSS, P.   *The John Tradescants, Gardeners to the Rose and Lily Queen* Peter Owen, 1984

11. DAWTREY, DREWITT   *The Romance of the Apothecaries Garden, Chelsea* Chapman Dodd, 1922

12. HAGBERG, KNUT   *Carl Linnaeus* Jonathan Cape, 1952

13. TAYLOR, GEOFFREY   *Some Nineteenth Century Gardeners* Skeffington, 1951

14. HOWE, BEA   *Lady with Green Fingers* Country Life, 1961

15. BUCHAN, URSULA *An Anthology of Garden Writing* Croom Helm, 1986

16. MASSINGHAM, BETTY *Miss Jekyll. Portrait of a Great Gardener* Country Life, 1966

17. JEKYLL, GERTRUDE   *Wood & Garden* Antique Collectors' Club, 1981

18. SCOTT JAMES, ANNE   *Sissinghurst, The Making of a Garden* Michael Joseph, 1975

SELECT BIBLIOGRAPHY

ALLEN, MEA *Plants that Changed our Gardens* David and Charles, 1974

ANDERSON, A. W. *The Coming of the Flowers* Williams and Norgate, 1950

BEAN, W. J. *Trees and Shrubs Hardy in the British Isles* (8th edn.) John Murray, 1976

BOWLES, E. A. *My Garden in Spring* T. C. & E. C. Jack, 1914

BOWLES, E. A. *A Handbook of Crocus and Colchicum for Gardeners* John Lane, The Bodley Head, 1952

BRAY, LYS DE *Manual of Old-Fashioned Flowers* Oxford Illustrated Press, 1984

BRICKELL, C. D. & SHARMAN, F. *The Vanishing Garden* John Murray, 1986

BRICKELL, C. D. *Journal of the Royal Horticultural Society,* Volume 89, Part 1, January, 1964 (pp. 19-22)

COATS, ALICE *Flowers and their Histories* Hulton Press, 1956

COATS, ALICE *Garden Shrubs and their Histories* Vista Books, 1963

COATS, ALICE *The Quest for Plants* Studio Vista, 1969

FISHER, JOHN *The Origins of Garden Plants* Constable, 1982

GENDERS, ROY *The Cottage Garden* Pelham Books, 1983

GIBSON, MICHAEL *Growing Roses* Croom Helm, 1984

GORER, RICHARD *The Growth of Gardens* Faber, 1978

GORER, RICHARD *The Development of Garden Flowers* Eyre and Spottiswoode, 1970

HADFIELD, MILES *A History of British Gardening* Murray, 1979

HYAMS, E. & MCCQUITTY, W. *Great Botanical Gardens of the World* Bloomsbury 1985

INGRAM, COLLINGWOOD *Ornamental Cherries* Country Life, 1948

JEKYLL, GERTRUDE *Wood and Garden* Antique Collectors' Club, 1981

JEKYLL, G. & WEAVER, L. *Gardens for Small Country Houses* Antique Collectors' Club, 1981

KINGDON-WARD, FRANK *Pilgrimage for Plants* Harrap, 1960

LEMMON, KENNETH *The Golden Age of Plant Hunters* Phoenix, 1968

LOUDON, JANE *The Ladies' Flower Garden of Ornamental Perennials* 1844

LOUDON, JANE *British Wild Flowers* Loudon, 1859

LOUDON, J. C. *Encyclopaedia of Gardening (1822 edn.)* Loudon, 1822

LYTE, CHARLES *The Plant Hunters* Orbis, 1983

LYTE, CHARLES *Sir Joseph Banks* David and Charles, 1980

PLOWDEN, C. CHICHELEY *A Manual of Plant Names* Allen and Unwin, 1970

PULTENEY, R. *Botany in England* T. Cadell, 1790

READER'S DIGEST *The Gardening Year* (ed. Roy Hay), 1978

READER'S DIGEST *Encyclopaedia of Garden Plants and Flowers* (ed. Roy Hay), 1971

ROBINSON, WILLIAM *The English Flower Garden* John Murray, 1898

ROYAL HORTICULTURAL SOCIETY *The Dictionary of Gardening* (ed. Fred J. Chittenden and Patrick M. Synge) Oxford University Press, 1974

Royal Horticultural Society's Encyclopaedia of Practical Gardening Mitchell Beazley, 1980

SACKVILLE-WEST, V. *V. Sackville-West's Garden Book* Michael Joseph, 1968

TAYLOR, JANE *The Plantsman,* Volume 7, Part 3, December, 1985 (pp. 129-160)

TERGIT, GABRIELE *Flowers through the Ages* Wolff, 1961

SPECIALIST NURSERIES

DAVID AUSTIN ROSES (Roses, paeonies and unusual perennials) Bowling Green Lane, Albrighton, Wolverhampton WV7 8HB.

PETER BEALES ROSES (Historic and modern roses) Intwood Nurseries, Swardeston, Norwich, Norfolk NR14 8EA.

BRESSINGHAM GARDENS (Plants of all kinds, garden) Bressingham, Diss, Norfolk IP22 2AB.

BROADLEIGH GARDENS (Small and unusual bulbs) Barr House, Bishops Hull, Taunton, Somerset.

BETH CHATTO, UNUSUAL PLANTS (Rare perennials, good garden) White Barn House, Elmstead Market, Colchester, Essex CO7 7DB.

HILLIER NURSERIES LTD (Shrubs and trees) Ampfield House, Ampfield, Romsey, Hampshire SO5 9PA.

HOPLEY'S PLANTS (Plants of all kinds, garden) High Street, Much Hadham, Hertfordshire SG10 6BU.

PARADISE CENTRE (Bulbs, herbaceous plants) Twinstead Road, Lamarsh, Bures, Suffolk CO8 5EX.

RAMPARTS NURSERY (Silver foliage plants) Bakers Lane, Colchester, Essex CO4 5BB.

STYDD NURSERIES (Tender perennials, roses) Stoneygate Lane, Ribchester, Nr. Preston, Lancashire.

SHERRARDS (Shrubs and trees) The Garden Centre, Wantage Road, Donnington, Newbury, Berkshire RG16 9BE.

TREASURES OF TENBURY LTD (Good garden) Burford House Gardens, Tenbury Wells, Worcestershire WR15 8HQ.

OTHER USEFUL ADDRESSES

THE ALPINE GARDEN SOCIETY, Lye End Link, St. Johns, Woking, Surrey.

THE CHELSEA PHYSIC GARDEN, 66 Royal Hospital Road, London, SW3 4HS.

THE HARDY PLANT SOCIETY, 10 St. Barnabas Road, Emmer Green, Caversham, Berkshire RG4 8RA.

THE NATIONAL COUNCIL FOR THE CONSERVATION OF PLANTS AND GARDENS, Wisley Garden, Woking, Surrey GU23 6QB.

THE NATIONAL GARDENS SCHEME, 57 Lower Belgrave Street, London SW1W 0LR.

THE NATIONAL TRUST, 36 Queen Anne's Gate, London SW1H 9AS.

THE NATIONAL TRUST FOR SCOTLAND, 5 Charlotte Square, Edinburgh EH2 4DU.

THE ROYAL HORTICULTURAL SOCIETY, Vincent Square, London SW1P 2PE.

A Selection of Gardens Open to the Public

The National Trust owns many of the gardens open to the public in Britain. Details can be obtained from them at the address given above.

There are many other excellent gardens, some privately owned, of which only a tiny selection is listed below:

CAREBY MANOR GARDENS, Stamford, Lincolnshire.

GREAT COMP CHARITABLE TRUST, Borough Green, Kent.

HATFIELD HOUSE, Hatfield, Hertfordshire.

HODNETT HALL GARDENS, Market Drayton, Shropshire.

HERGEST CROFT GARDENS, Kington, Herefordshire.

HYDE HALL GARDEN TRUST, Rettendon, Chelmsford, Essex.

JENKYN PLACE, Bentley, Hampshire.

KIFSTGATE COURT, Chipping Campden, Gloucestershire.

THE GARDEN HOUSE, Buckland Monachorum, Yelverton, Devon.

ROSEMOOR GARDEN TRUST, Torrington, Devon.

Index

Page numbers in *italic* refer to illustrations and captions.

A

Acanthus mollis, 54, *55*
    A.m. latifolius, 54
    A. spinosus, 54
Acer capillipes, 56
    A. davidii, 8, 56, *57*
    A. pennsylvanicum, 56
    A. rufinerve, 56
Agapanthus, 29
    A. africanus, 58, *59*
    A. Headbourne Hybrids, 58
    A. umbellatus, *34*, 58
Aiton, William, 32
Alchemilla mollis 46
Allium, 28
    A. albopilosum, 46
Allwood, Montagu, 86
Amelanchier ovalis, 22
Anemone nemorosa 'Allenii', 60
    A. robinsoniana, 60, *61*
anemones, *27*, 43
Angel's Fishing Rods, 88, *89*
Aquilegia canadensis, 62, *62*
    A.c. 'Nana', 62
    A. flavescens, 62
Arbutus unedo, 64, *65*
Ashmole, Elias, 22
asphodels, 28
Aster novae-angliae, 41, 66, *67*
Atragene alpina, 76, *77*
auriculas, 50
Austin, George, 70
Australian bottle-brush, 70, *71*

B

Backhouse, 88
Bacon, Frances, 18
Balloon Flower, 124, *125*
Banks, Sir Joseph, 34, 36, 70
Bear's Breeches, 54, *55*
Bear's Ears, 128, *129*
blackcurrants, 21
Bloom, Alan, 102
Boraginaceae, *19*
bottle-brushes, 70, *71*
Bowles, E.A., 47-9, *47*, 60, 78, 92, 94, 150
box, 11, 16
Brompton Park Nursery, 29
Brown, Lancelot Capability, 33, 39
Buckingham, Duke of, 21
Buddle, Adam, 68
Buddleia, 8
    B. alternifolia, 68
    B. davidii, 68, *69*
    B. fallowiana, 68
    B. globosa, 68
Buonaiuti, 82
Busbecq, Ogier Ghiselin de, 122, 148
Butternut, 22

C

Callimachus, 54
Callistemon spp., 70, *71*
    C. linearis, 70
    C. salignus, 70
Camellia japonica, 72-3, *73*
    C. saluenensis, 72
    C.× williamsii, 72
campion, *13*

carrots, 12
Cecil family, 18
Cedar of Lebanon, 31
Cels, 106
Chaenomeles japonica, 74
    C. speciosa, 74, *75*
    C.× superba, 74
Chatsworth, 38
Chaundler, 18
Cheiranthus 'Bowles Mauve', 47
Chelsea Physic Garden, 30-2, *33*
cherries, 22
Chinese Bell-flower, 124, *125*
Christmas Rose, 12, 98-9, *99*
Clematis, 41
    C. alpina, 76, *77*
    C. 'Ernest Markham', 50
    C. 'Markham's Pink', 50
Clovis I, 101
Clusius (Jules Charles L'Ecluse), 14-15, *15*, 92, 128, 148
Colchicum, 28, 48
Columbine, 62, *63*
comfrey, 46
Cook, Captain James, 70
Cotinus cogyggria, 22
cotton, 32
Cotton Lavender, *19*
Crimson Flag, 142, *143*
Crocus, 13, 48
    C. biflorus, 78
    C. chrysanthus, 78-9, *79*
    C.c. pallidus 'E.A. Bowles', 78
Crown Imperial, 92-3, *93*
cycads, 50

Cyclamen, 9, 28
  C. hederifolium, 10
    C.h. 'Bowles Apollo', 47
  C. persicum, 80-1, *81*
Cytisus battandieri, 48
D
daffodils, 48, 114
Dahl, 82
Dahlia coccinea, 82, *83*
  D. pinnata, 82
  D. rosea, 82
  D. collina, 48, 84, *85*
  D.× hybrida, 84
  D. mezereum, 84
  D. odora, 84
David, Père Armand, 56, 68
Davidia involucrata, 56
David's Maple, 56, *57*
Delavay, Abbé, 108
Dianthus× allwoodii, 86
  D. barbatus, 17
  D. caryophyllus, 86
  D. plumarius, 86-7, *87*
Dierama pulcherrimum, 48, 88, *89*
Digges, Sir Dudley, 20, 21
Digitalis ambigua, 90
  D.× mertonensis, 90
  D. purpurea, 90, *91*
Dioscorides, 8-10
Dittany, 46
Dodoens, Rembert, 17
Douglas, David, 96
Duplessis, Marie, 72
Dusty Miller, 128, *129*
E
Egypt, ancient, *9*
Ellacombe, Henry, 48
Eranthis hiemalis, 94
Erinus, 41
Evelyn, John, 24-6, *26*, *27*, 54
F
Farrer, Reginald, 47, 48
Fever Bush, *96*
Figs, 8
Finger-flower, 90, *91*
Flowering Currant, 96
Forest, George, 56
Forget-Me-Not, 110
Foxglove, 90, *91*
Fritillaria imperialis, 92-3, *93*
  F.i. maxima lutea, 92
  F.i. maxima rubra, 92
Fulmitory, 41
G
Galanthus, 48
  G. elwesii, 47
  G. nivalis, 94-5, *95*
    G.n. flore pleno, 94
    G.n. scharlokii, 94
    G.n. viridapicus, 94
  G. plicatus, 47
Galpin's Red Hot Poker, 102, *103*
Gambier, Lord, 152
Gardener, Ion, 11-12
Garlic, 12
Garry, Nicholas, 96

Garrya elliptica, 96, 97
Geranium, 50
Gerard, John, 16, *16*, 17, 18, 86, 92, 94, 98, 114, 118, 122, 128
Germander Speedwell, 110
Gillyflowers, 24
Gladioli, 28
Goodyer, John, 8
Granny's Bonnet, 62, *63*
Grapes, 8, 20, 22, 24
Gravetye Manor, *39*, 41
Grieve, James, 152
H
Hampton Court, 14, 16, 29, *30*, *49*
Handkerchief tree, 56
Hanmer, Sir Thomas, 23-4, 26, 146
Hardwicke, Thomas, 136
Hatfield House, 18, 20, *21*, 22
Hawthorn, 13
Heartsease, 152
Hegarty, Blanche, 142
Helleborus, 43
  H. foetidus, 98
  H. niger, 12, 98-9, *99*
  H. orientalis, 98
  H. viridis, 98
Henry, Augustine, 68
Hepatica, 14, 22
Hestercombe, *42*
Highclere, 136
Hippocrates, 126
Hogelande, Johan de, 15
Holland, Lady, 82
Holly, 24, 26
Hornbean, 29
I
Indian corn, *16*
Ingram, Collingwood, 132
Iris, 12, 14, *27*, 46
  I. florentina, 9
  I. germanica, 101-2, *102*
Iver, 152
J
Jekyll, Gertrude, 41-4, *41*, *42*, *43*, 98, 102, 108, 140
Joseph Rock's Rowan, 144
Juglans cinerea, 22
K
Kaffir Lily, 132, *143*
Kamel, George Joseph, 72
Kew Gardens, *37*
Kingdom-Ward, Frank, 47, 56, 108, 130
Kinlet, 26
Kniphofia galpinii, 102, *103*
  K. macowanii, 102
  K. triangularis, 102
  K. uvaria, 102
Knole, 45
L
Lady Tulip, 148, *149*
Larch, 20
Laurel, 24
Laurustinus, 150, *151*
Lawrence, Sir Trevor, 50
Le Nôtre, André, 23

Lenné, Herr, 106
Lent Lily, 114, *115*
Lenten Rose, 98
Levant Company, 126
Lilac, 122
Lilium candidum, 13, 104, *105*
  L.c. cernum, 104
Limes, pleached, 46
Lindley, John, 37
Linnaeus (Carl von Linné), 32, *32*, 33-4, *34*, 62, 120
Liverwort, *13*, 14
L'Obel, Matthias de, 20
Lobelia, 20
Lockar, Roger, 30
London, George, 29-30, 31, 120
London and Wise, 30, *49*, 58
London Flag, 100-1, *101*
London Plane, 22, 126
Loudon, Jane, 36, 37-8, *38*, 41
Loudon, John Claudius, 35-8, *35*, 37-8, 41, 82, 106
Louis VII, King of France, 99
Love apples, 17
Ludlow, Frank, 108
Lutyens, Sir Edwin, *41*, 42, *42*, 43-4
M
Madonna Lily, 13, 104, *105*
Magnol, Pierre, 106
Magnolia denudata, 106
  M. liliiflora, 106
  M.× soulangiana, 106-7, *107*
Mahonia aquifolium, 96
Main, James, 74
Maize, *16*
Maries, Charlies, 124
Markham, Ernest, 50
Martagon Lilies, 22
Meconopsis betonicifolia, 108
  M. grandis, 108, *109*
  M. napaulensis, 108
Medlars, 28
Menzies, Archibald, 96
Miller, Philip, 30-2, 33, *33*, 64, 80, 134
Mock Orange, 122
Monardes, Nicholas, 15
More, Sir Thomas, 14
Morning Glory, 140
Morris, William, 42
Mountain Cowslip, 128, *129*
Mullein, *28*
Myosotis alpestris, 110, *111*
Myrtle, 112
Myrtus communis, 112, *113*
N
Narcissus poeticus, 114
  N. pseudonarcissus, 114, *115*
Neckham, Alexander, 12
Nerine bowdenii, 25
  N. sariensis, 25
Nicholson, Harold, 44-7, *45*
Nicotiana affinis, 116
  N. alata, 116, *117*
Norton, John, 17

O
Oatlands, 22
Opium poppy, 12, 120, *120*
Oranges, 23
Oregon Grape, 96
Oriental Plane, 126, *127*
Oriental Sycamore, 126, *127*
Orris Root, 9
Oxford Botanic Gardens, 28, *28*, 33
P
Paeon, 118
Paeonia officinalis, 118, *119*
   P. suffruticosa 'Rocks Variety',
    144
paeonies, 12, 27, 118
Paisley weavers, 50, 82, 86, 128
Palmer, Lewis, 58
Palms, 50
Pansies, 50, 152, *153*
Papaver orientale, 120
   P. somniferum, 120, *120*
    P.s. paeoniaeflorum, 120
Parkinson, John, 18-20, *19, 20*, 22,
   27, 64, 86, 118, 128, 134,
   146
Passiflora, 20
Passion flower, *20*
Paxton, Joseph, 38, 40, 82
Peaches, 24
Pears, 13, 24, 27
Peas, *12*
Pemberton, J.H., 50
Perry, Amos, 102
Persian Lilac, 146, *147*
Philadelphus, 148
   P. coronarius, 122, *123*
Phipps, *41*
Pineapples, 18
Pinks, *24*, 86-7, *87*
Plantanus× acerifolia, 126
   P. occidentalis, 126
   P. orientalis, 126, *126*
Platycodon grandiflorus, 124, *125*
   P.g. var. *mariesii*, 124
Pliny the Elder, 10-11, *10*, 13
Pliny the Younger, 11
Plunkenet, 58
Polyanthus, 43, 50
Pomegranate, 8, 134, *135*
Potato, 17
Prain, David, 108
Prickly Pears, 18
Primroses, 28, 43, 46-7
Primula auricula, 128, *129*
   P. florindae, 130, *130*
   P.× pubescens, 128
   P. rubra, 128
Prunus serrulata var. *spontanea*,
   132
   P. 'Tai-Haku', 132, *133*
Pseudotsuga taxifolia, 96
Pulmonaria, 46
Punica granatum, 134, *135*
Purdom, William, 47
Purple Flag, 100-1, *101*
Pyrame de Candolle, Augustin, 126

Q
Quince, 27
Quinine Bush, 96, *96*
R
Raleigh, Sir Walter, 116
Ranunculus, *27*, 50
Rea, John, 26-8, *27*, 128
Regents Park, 39-40
Repton, Humphry, 33, 39
Rhododendron× altaclarense, 136
   R. arboreum, 136, *137*
    R.a. campbelliae, 136
    R.a. cinnamomeum, 136
   R. catawbiense, 136
   R. hirsutum, 136
   R. ponticum, 136
Ribes sanguineum, 96
Robin, Vespasien, 22
Robinson, William, 39-42, *39, 40*, 54,
   60, 66, 102, 140, 152
Robinson's Windflower, 60, *61*
Rock, Joseph, 144
Rosa 'Buff Beauty', 50
   R. gallica 'Versicolor', 14, 138,
    139
   R. longicuspis, 46
   R. 'Penelope', 50
   R. 'Rose de Maures', 46
   R. 'Sissinghurst Castle', 46
'Rosa Mundi', 14, 138, *139*
Rose, John, 29
Rosemary, 11, 16
Roses, *12*, 13, *13*, 14, 27, 46, 50, 138
Rouen Lilac, 146
Runner beans, 22
S
Sackville-West, Vita, 44-7, *44, 45*, 98
Saffron, 13
Salisbury, Lord, 20, 22
Salvia patens, 140, *141*
Sayes Court, 26
Schizostylus coccinea, 142, *143*
Sea Catchfly, 14
Sherriff, George, 108
Silene vulgaris, 14
Silk Tassel Bush, *96*
Sissinghurst Castle, 44, 45-7, *45, 46*
Slinger, Leslie, 88, 102
Slinger, W., 88, 102
Sloane, Sir Hans, 31
Smilacina racemosa, 22
Smith, J.E., 34
Smith, James, 70
Smoke Bush Lilac, 22
Snakebark Maple, 56
Snowdrop, 48, 94-5, *95*
Sorbus, 28
   S. 'Joseph Rock', 144, *145*
Soulange-Bodin, M., 106
Soulié, Père, 68
Sowbread, 80
Stern, Sir Frederick, 48
Strawberries, 21
Strawberry Tree, 64, *65*
Stuart, Dr, 152
Sturtevant, 18

Sweet Cicely, 41
Sweet William, 17
Switzer, Christoph, 19
Syringa× chinensis, 146
   S. persica, 146, *147*
   S. vulgaris, 122
T
Tasman, 70
Theobalds, 17
Theophrastus, 94
Thompson, 152
Tobacco Plant, 116, *117*
Tomatoes, 17
Torch Lily, 102, *102*
Tradescant, John the Elder, 18, 20-1,
   21, 22, 62, 86, 146
Tradescant, John the Younger, 20, 22-
   3, *22, 23*
Tradescantia, 22
   T. virginiana, *23*
Tulipa, 14, 24, 27, 28, *30*, 49, 50
   T. clusiana, 14, *15, 26*, 148-9, *149*
Turner, William, 54, 64, 114, 118, 126
Tusser, Thomas, 16, 18
V
Van Tubergen, 78
Vancouver, George, 96
Veitch nursery, 56, 124
Veronica, *28*
Viburnum tinus, 150, *150*
Vilmorin, 68
Viola 'Bowles Black', 47
   V. cornuta, 152
   V. lutea, 152
   V. tricolor, 152-3, *153*
   V.× williamsii, 152
   V.× wittrockiana, 152
violets, *13*, 14
W
Wand Flower, 88, *89*
Watts, John, 31
White Mulberry, 20
Wild Bay, 150, *151*
Wild Daffodil, 114, *115*
Wilks, William, 50
Williams, J.C., 50
Wilson, E.H., 47, 56
Winter Aconite, 94
Wise, Henry, 29
Witch's Thimble, 90, *91*
Withering, William, 90
Wotton, Sir Edward, 90
Y
Young, Messrs, 106

ACKNOWLEDGEMENTS

I should like to acknowledge gratefully the assistance of Dr. Brent Elliott and his staff at the Royal Horticultural Society's Lindley Library, and the staff of the University Library, Cambridge. I should also like to thank Toby Buchan for answering arcane queries so readily and my husband, Charles Wide, for his constant support.

URSULA BUCHAN

My grateful thanks to Dr. Brent Elliott and his assistant, Barbara Collecott, at the Royal Horticultural Society's Lindley Library for their invaluable assistance. I should also like to thank my wife, Rosamund, for her support, and apologise to my children for neglecting them during the preparation of this book.

NIGEL COLBORN

PICTURE CREDITS

p.9 By courtesy of the Trustees of the British Museum
p.10 Victoria and Albert Museum/E.T. Archive
p.11 Scala
p.12 E.T. Archive
p.13 National Gallery
p.14 Royal Horticultural Society/photo Eileen Tweedy
p.16 Bodleian Library
p.19 Victoria and Albert Museum/photo Eileen Tweedy
p.20 Victoria and Albert Museum/photo Eileen Tweedy
p.21 John Bethell
p.22 Ashmolean Museum
p.23 Royal Horticultural Society/photo Eileen Tweedy
p.24 Royal Horticultural Society/photo Eileen Tweedy
p.24 Hunt Institute for Botanical Documentation/ Carnegie Mellon University, Pittsburgh, Pa
p.25 Royal Horticultural Society/photo Eileen Tweedy
p.26 Royal Horticultural Society/photo Eileen Tweedy
p.26 Royal Society (portrait of Evelyn)
p.27 Royal Botanic Gardens, Kew
p.28 John Bethell
p.29 By courtesy of Her Majesty the Queen
p.30 E.T. Archive
p.32 E.T. Archive
p.33 E.T. Archive
p.34 E.T. Archive
p.35 Royal Horticultural Society/photo Eileen Tweedy
p.37 John Bethell
p.38 Royal Horticultural Society/photo Eileen Tweedy
p.39 Peter Herbert, Gravetye Manor
p.40 Peter Herbert, Gravetye Manor
p.41 Royal Institute of British Architects
p.42 Country Life
p.43 National Portrait Gallery
p.44 Sothebys
p.45 John Bethell
p.46 Edwin Smith
p.47 Royal Horticultural Society/photo Eileen Tweedy
p.49 John Bethell